Advances in Spatial Science

Springer
Berlin
Heidelberg
New York
Barcelona
Budapest
Hong Kong
London
Milan
Paris
Santa Clara
Singapore
Tokyo

Titles in the Series

Christoforo S. Bertuglia, Manfred M. Fischer
and *Giorgio Preto (Eds.)*
Technological Change, Economic Development and Space
XVI, 354 pages. 1995
ISBN 3-540-59288-1

Harry Coccossis and *Peter Nijkamp (Eds.)*
Overcoming Isolation
Information and Transportion
Networks in Development Strategies for Peripheral Areas
VIII, 272 pages. 1995
ISBN 3-540-59423-X

Luc Anselin and *Raymond J.G.M Florax (Eds.)*
New Directions in Spatial Econometrics
XIX, 420 pages. 1995
ISBN 3-540-60020-5

Heikki Eskelinen and *Folke Snickars (Eds.)*
Competitive European Peripheries
VIII, 271 pages. 1995
ISBN 3-540-60211-9

Jeroen C.J.M. van den Bergh, Peter Nijkamp
and *Piet Rietveld (Eds.)*
Recent Advances in Spatial Equilibrium Modelling
VIII, 392 pages. 1996
ISBN 3-540-60708-0

Peter Nijkamp · Gerard Pepping
David Banister

Telematics and Transport Behaviour

In Association with:
Yorgos Argyrakos, Paolo Camara, Richard Emmerink
Francesco Filippi, Maria Giaoutzi, Maarten Hilferink
Francesca Kenny, Massimo Marciani, Hans Ouwersloot
Kostas Petrakis, Laurie Pickup, Aura Reggiani

With 54 Figures
and 54 Tables

Springer

Professor Dr. Peter Nijkamp
Drs. Gerard Pepping

Free University
Department of Spatial Economics
De Boelelaan 1105
NL-1081 HV Amsterdam, The Netherlands

Professor Dr. David Banister
University College London
The Bartlett School of Planning
Wates House, 22 Gordon Street
UK-London WC1H OQB, United Kingdom

Cataloging-in-Publication Data applied for

Die Deutsche Bibliothek - CIP-Einheitsaufnahme

Nijkamp, Peter:
Telematics and transport behaviour : with 54 tables / Peter Nijkamp ; Gerard Pepping ; David Banister. - Berlin ; Heidelberg ; New York ; Barcelona ; Budapest ; Hong Kong ; London ; Milan ; Paris ; Santa Clara ; Singapore ; Tokyo : Springer, 1996
(Advances in spatial science)
ISBN-13: 978-3-642-80141-9 e-ISBN-13: 978-3-642-80139-6
DOI: 10.1007/ 978-3-642-80139-6
NE: Pepping, Gerard:; Banister, David:

Softcover reprint of the hardcover 1st edition 1996

SPIN: 10516891 42/2202-5 4 3 2 1 0 - Printed on acid-free paper

Preface

Global villages, wired societies and electronic highways are the contemporary images of the spatial organization of our society in the information age. Such concepts all refer to the spatially structuring impact of the adoption and dissemination of the modern information and telecommunication technology. It does not require much fantasy to depict a future society where interaction and communication are mainly driven by sophisticated information networks. One of the fields where the modern information and telecommunication sector may exert a significant impact is transportation.

Transportation all over the world has become a strategic activity connecting people, cities and regions. It is in fact the red thread connecting all nodes of a modern network economy, at both international and local levels. At the same time however, the transport sector is facing severe impediments which hamper the development potential of this network economy. Examples of externalities preventing our society from reaping the full fruits of transportation are social costs from traffic insecurity, environmental pollution and traffic congestion.

Policies to cope with such negative externalities of transport may be manifold in nature, for example, market-based incentives (road pricing, e.g.), regulatory measures, or technological developments (vehicle technology, improvement of certain modes etc.). One of the new technology advances aiming at making the transport system more efficient, is the introduction of advanced transport telecommunication and information technologies or *transport telematics*. This has become a rather popular policy option in many countries, as it tries to create a smoother operation of transport flows without requiring a drastic expansion of physical infrastructure. It is often - uncritically - assumed by public bodies and the industry that such new systems will act as a 'deus ex machina'.

Transport telematics is not a uniform technology, but has a great variety of appearances and applications, ranging e.g. from radio information to interactive route guidance. It can be used at various geographical scales, from local (e.g. parking information) to global (e.g. electronic data interchange). The use of telematics in the transport sector has called for much attention in the United States of America, Japan and Europe, and it has generated a great diversity of research initiatives, as it has become evident that transport telematics is not 'manna from heaven' which can be used as an automatic medicine against the evils of the transport sector.

There is a need to take a more critical perspective in analyzing the potential impacts of modern transport telematics. Therefore, we will abandon a purely engineering viewpoint ('technology push' idea) on the future impacts of transport telematics and focus in particular on the market side (i.e., the users and actors in this field). Such a behavioural approach is necessary to restore the balance against futuristic expectations on the introduction of telematics in transport.

The present book offers both a behavioural framework for analyzing transport telematics and brings together a series of studies on transport telematics and its implications for spatial behaviour of users in different fields (e.g. private car drivers, users of public transport, fleet operators). The book aims to offer a state of art survey of many applications of transport telematics, with a particular view on behavioural and policy responses. It is unique in that for the first time in the recent history of telematics, a comprehensive analysis is given of the fruits of this new technology. Hence, it contains many case studies on human behaviour and transport telematics.

Most studies presented in this book are the offspring of the European DRIVE programme on transport telematics. The empirical results have in particular been achieved in the framework of BATT (Behaviour and Advanced Transport Telematics) and MARTA (Monitoring Attitudes towards Road Transport Automation) projects in the DRIVE programme. These projects were a multinational undertaking in which several fellow-researchers from different countries were involved. Their names can be found on the front cover page of this book. They have to be thanked for their creative contribution to the study of behavioural consequences of transport telematics in Europe. Their collegiality has been a rich source of scientific inspiration and social enjoyment.

The authors also wish to thank the Tinbergen Institute in Amsterdam for acting as a host during the writing process of the book. Our thanks also go to the VSB Fonds (Utrecht) which has offered a grant for research in the field of 'Environment and Transport'.

In the final stage we received useful comments from Manfred Fischer (Vienna). The text editing has been a responsibility of Dianne Biederberg (Contact Europe, Amsterdam). Finally, we acknowledge also the creative support of Marianne C.P. Bopp (Springer-Verlag) who in the production stage of this book convincingly demonstrated the social potential of telematics in keeping us on the same pathway.

Amsterdam/London, November 1995

Peter Nijkamp, Gerard Pepping, David Banister

Contents

PART A

ROAD TRANSPORT TELEMATICS: SUPPLY AND DEMAND

1 Mobility and Telematics

1.1 Introduction

Spatial mobility has become one of the characteristic features of a modern economy. The so-called 'homo mobilis' has become a dominant actor whose ambition seems to be to increase his or her action space at ever increasing speeds. Time and space preferences appear to be two closely interwoven phenomena which determine the ongoing rise in spatial interactions. Europe is offering an illustrative example of this world-wide trend.

Following the completion of the internal market, the further political integration towards a European Union and the expansion towards a Europe of the Fifteen, Europe is increasingly moving towards a *network economy* in which physical flows of people and goods constitute the main connecting elements. This drive towards a European network society implies also that nodes and edges will become critical infrastructural elements for a rise in general economic performance in Europe, as well as for the welfare position of its constituent regions (cf. Capello 1994). This means that sufficient concern is warranted for transport conditions in both European *activity centres* (e.g. metropolitan areas, cities) and on the *connecting links* (e.g. road and rail infrastructure). If Europe wishes to reap the fruits of its integration, then the network infrastructure should not form an obstacle but an opportunity for competitive behaviour of all actors in the European space (see Batten et al. 1994; Johansson et al. 1994; Nijkamp 1994).

At the same time, it has to be recognized that network operations in the European transport sector face severe problems as a result of *traffic congestion*, *environmental decay* (ranging from local quality of life conditions to global climatological change) and a *lack of safety*. The social costs of the transport sector in Europe appear to be unacceptably high, up to a level of some 4 percent of GDP in various countries (Himanen and Nijkamp 1992; Verhoef 1994). Thus the European transport sector is confronted with severe capacity constraints of all kinds. Further average growth figures of mobility in Europe of 50% and more in the next 20 years are foreseen (Bovy et al. 1993). Such further growth may be prohibitive if these problems are not solved in the near future. Conventional transport policy is apparently unable to cope with Europe's growing demand for more mobility.

A variety of policies can be imagined to alleviate some of the problems on European infrastructure networks, ranging from market principles (e.g., road pricing or cordon pricing) to institutional solutions (e.g., prohibitions of certain types of transport) and technology solutions (e.g., zero emission cars). In recent years, much interest has been expressed in many countries to cope with the capacity problems on European infrastructure and to increase the efficiency or attractiveness of certain transport modes by employing new technological opportunities offered by the information and communication technology, in particular telecommunications. This blend of information technology and telecommunication technology in the transport sector is usually called *transport telematics*. Sometimes one also speaks of Advanced Transport Telematics (ATT). Transport telematics is increasingly seen as a new possibility for improving the performance of transport systems while respecting at the same time conditions imposed by environmental and safety goals. In this respect, telematics is often advocated - by both governments and the industry - as a powerful tool for coping with future transport problems in the emerging European network economy. However, the mutual relationship between telecommunications and transport is complex and includes various kinds of different impacts (Mokhtarian 1990 and Capello 1994). Also there is not a single telematics technology, but a wide variety of telematics possibilities and some main categories will be discussed in the next section.

1.2 Tele Applications in the Transport Sector

The use of telematics in the transport sector has different appearances and concerns different modes.

In the first phase, one may distinguish *the provision and use of information* by means of telecommunication to trip-makers in order to increase the efficiency and reliability of transport operations. Electronic data interchange (EDI) in the freight transport sector is a good example. The same holds for automatic debiting systems for parking or road pricing. In this case, telematics does not necessarily affect transport behaviour (in terms of route choice, trip scheduling, departure time), but it serves to increase the performance of transportation. This is more or less an a-spatial use of telematics in the transport sector.

Next, we may mention various telematics technologies which may have an *immediate day-by-day consequence for transportation behaviour*. Examples are route guidance, variable message signs or radio data information. Such uses of telematics influence spatial behaviour of trip-makers, not only for car users and truck drivers but also for users of public transport.

And finally we may distinguish telematics applications which have a *structuring impact on mobility behaviour*. This may be found in tele-working, tele-commuting, tele-conferencing, tele-shopping etc., where trip-making (e.g., home-to-work) is influenced in terms of changes in commuting or shopping patterns (cf. Hamer et al. 1991). Another example concerns tele-centres which is a decentralized concentration of tele-working or tele-commuting (see e.g. Salomon 1990). In most cases the latter type of telematics use has a substituting impact on transport behaviour of individuals (cf. Nilles 1991), and the beneficial impacts may therefore be relatively large in comparison with the former two types of telematics use. Recent research shows that the potential of teleworking is considerable in terms of reduced trips and impacts on the environment (e.g. Mokhtarian et al. 1995; Henderson et al. 1995; Handy et al. 1995).

It is clear that the adoption of telematics technologies is a multi-faceted phenomenon, in which time preferences, individual motives, labour flexibility, financial sacrifices and many other driving forces play an important role. This also means that one cannot speak of *the* telematics technology, but of a wide range of technologies which may have to be tuned to the needs of individuals and corporate organisations. Thus, there is scope for a wide spectrum of scientific research, which should not only be technological in nature, but also - and increasingly important - *social science oriented*. Transport telematics is not a matter of technological sophistication in a technology-push view on the world, but a matter of social acceptance based on market-pull strategies. Consequently, research in the transport telematics sector should address problems of individual perception, social attitude, behaviour of firms and personal motives on many elements which are directly or indirectly related to transport behaviour. This will also be the spirit of this study on the social dimension of telematics applications. Consequently, behaviour of users of transport telematics will be given due attention with a particular view on the various actors involved in choosing and adopting different types of telematics applications.

1.3 Organization of the Book

The social dimension of transport telematics comprises two main elements: on the one hand the behavioral response of individual travellers in terms of their pre-trip and on-trip travel choices, and on the other hand the overall market potential of transport telematics. This book deals with both issues in a consistent manner, by combining theoretical and conceptual issues with practical cases and policy analyses. This emphasis on behavioral approaches leads to a logical division of the book into three parts (see Figure 1.1). The focus is entirely on the road transport sector, as it covers both passenger and freight transport applications.

In the first part, Part A, we will take a look at the conceptual issues related to the range of transport telematics technologies, the characteristics of the potential user market and road network features which play a dominant role. Part A starts with a detailed description of the full range of possible functions in transport logistics that can be addressed by transport telematics technologies, as well as specific product types (in various phases of development). The next chapter, Chapter 3, discusses in depth the characteristics of the potential user groups (or actors) of these technologies. It is set out how different motives in the case of different actors play critical roles for the formation of the market potential for telematics. In addition, conceptual approaches are presented which are used to investigate trip behaviour, the diffusion and the market potential of transport telematics. This chapter provides a link to Part B of the book, in which in more detail the impact of telematics in the case of classes of actors will be discussed. Part A ends with Chapter 4. This chapter contains a cross-regional analysis of road corridors with respect to features affecting the site-specific (public) interest of telematics functions. It aims to develop a method to systematically assess the telematics potential on different sites, given the large differences between different road networks in Europe.

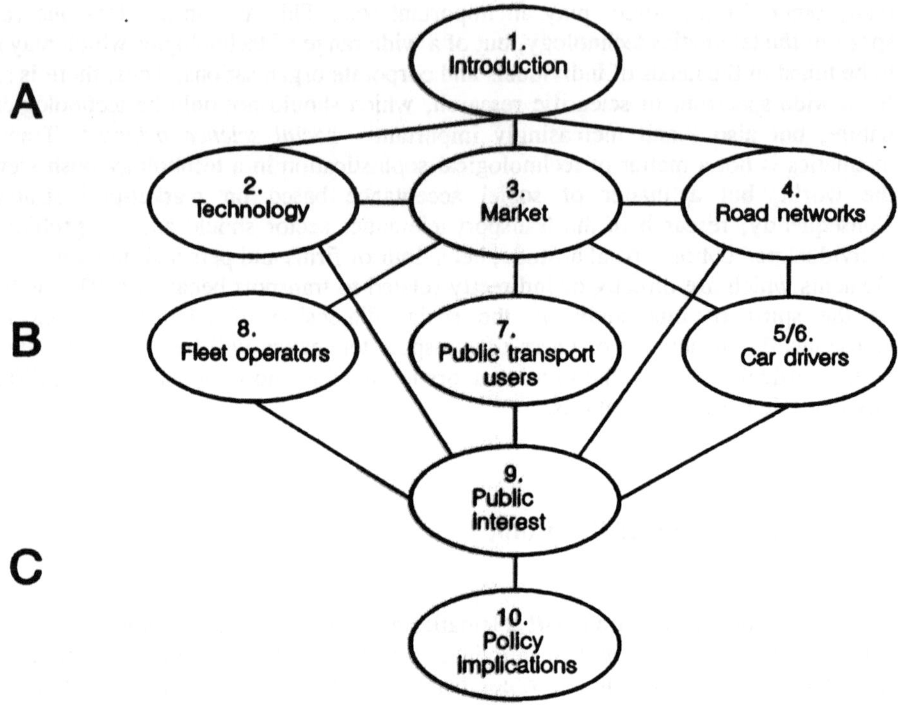

Fig. 1.1. Structure of the book.

Part B discusses, as already mentioned, the responses of different user groups to telematics. Three main actor groups on the user side are distinguished: *private car drivers*, *public transport users* and *road freight operators*. These groups will subsequently be dealt with in Chapters 5 and 6, 7 and 8, respectively.

In Part C finally, the future perspective for road transport telematics will be dealt with. It will start with an analysis of the public interest in transport telematics, based inter alia on an expert opinion approach (Chapter 9). Then, in Chapter 10 a synthesis of both the conceptual findings and the empirical evidence provided in this book will be presented, along with some clear recommendations for the future policy with regard to transport telematics.

2 Transport Functions of Telematics Technology Options

2.1 Supported Transport Functions and Technological Options

Transport telematics provides the transport sector with products and services of a varied and different form. It has no physical transport infrastructure as such; it is based on a parallel communications infrastructure which acts as a valuable 'enabling mechanism', transferring information to enhance the operation of the transport network. Transport telematics introduces potentially more effective and novel solutions for addressing the current and future transport problems, as has been outlined in the previous chapter. This is achieved by providing decision-makers, network actors and fleet operators with better information on the past and current state of the network, and by offering better information to the public on travel options and conditions when they plan or undertake trips.

This chapter has a two-fold aim. First, it identifies the range of possible *transport areas or functions* where transport telematics technologies could be implemented, and second, it provides a state-of-the-art review of *technological options* in the field of transport telematics, considering a wide range of telematics systems being in various stages of development. The range of transport functions for which telematics could be used determines in fact the potential size of the telematics market. Regarding the technological options discussed here, the discussion will focus primarily on hardware developments. Given these technical opportunities, we will in subsequent chapters address the potential of transport telematics for the various actors.

In general, a specific technology addresses more than one of such transport functions (depending on the way of classification). We may take as an example the provision of dynamic driver information by means of road-based information signs such as Variable Message Signs (VMS). Such systems are both tools for general traffic management (as messages are controlled by road network operators) and a source of traffic information to motorists who may all respond individually to these messages.

A comprehensive classification of transport functions where transport telematics could be implemented has been developed by OECD (1992) and CORD (1993). Based on this classification approach, we may distinguish seven major *functional transport areas*. In order of treatment within this section, these are:

1. travel and traffic information;
2. public transport management;
3. freight and fleet management;
4. road and traffic management;
5. demand management;
6. parking management;
7. driver assistance and cooperative driving.

This classification of main transport areas is ambiguous in the sense that some relevant transport functions may basically be classified under more than one of these main application areas. This is particular true for some functions in the areas of infrastructure management and traffic information.

A treatment of all existing technologies and technologies in development is a formidable task and is made even more difficult by the dynamic evolution of the transport telematics sector. Efforts to make a comprehensive inventory of technologies have been carried out in MARTA (1992, 1993b and 1995). Instead of presenting a comprehensive and detailed description of each technology, we will focus here on key application areas and technologies.

2.1.1 Travel and Traffic Information

The area of provision of travel and traffic information is closely related to the implementation of most other telematics functions: demand management, parking management, traffic management and public transport and freight operations. The functional area of travel information has applications for all kinds of travellers; it includes the provision of information before and during the trip to facilitate travel planning and travel bookings, as well as information to drivers about services whilst travelling. These kinds of information services have, in relation to the provision of traffic information (see below), a long duration time and a wide geographical dimension. The functional area of providing traffic information for vehicle drivers includes the provision of dynamic traffic and traffic-related information to drivers during the trip via either collective means (e.g. VMS) or individual in-vehicle terminals. This is generally characterized by short temporal and geographical spans, and is specific to the actual driving task and the location of the vehicle in the road network. All functions within these two areas are systematically presented in Tables 2.1 and 2.2.

The way in which travel information is disseminated to travellers can be either via classical means or via private and public interfaces like PC's or terminals. The latter form of information provision can be static or interactive, using either no, one- or bidirectional external communication links. The classical means to disseminate travel information are for example press and broadcasting by radio and TV (teletext). More advanced private terminals comprise personal portable units, in-vehicle units (see below under traffic information), and home and workplace terminals. Public terminals can be situated in urban centres (e.g. bus stations), all kinds of (motorway) service stations, etc. Travel information is supplied from multiple and diverse sources, e.g. public transport operators, rail operators and tourist organizations. These private and public systems have been available for some time and are being expanded to provide multi-modal travel information, to be real-time and to provide reservation facilities. The market of these systems is reliant on products being genuinely multi-service oriented.

Table 2.1. Travel Information Functions.

- Travel planning
Provision of information necessary for either single-mode or multi-mode travel planning, business and private - prior to the trip being made.

Transport mode selection/planning
Provision of information on scheduling of different transport modes, including static and actual time tables and tariffs.

Route planning
Provision of information on alternative routes, depending on user preferences. These preferences include the choice of single mode or multi-modal travel, constraints such as timing, expenditures and parameters associated with the selected mode(s) of travel.

Tourist information
Provision of tourist information like information about attractions, accommodation and other places of interest.

Services information
Information on service facilities such as garages, petrol stations, banks, post offices, hospital and other facilities.

Facilities booking
Procedures to make and pay for a booking.

- Static route information
Provision of information on static characteristics of roads and facilities along routes.

Road characteristics
Provision of information on geometry of roads, limitations of use such as height and weight, one-way streets etc.

Parking facilities
Provision of information on location of parking facilities along routes.

Modal interchange points
Provision of information on modal interchange points such as Park & Ride.

Individual map updating
Updating of individual road maps (manual or automatic).

- Personal communications

Emergency paging
Allowance of individuals to be paged for urgent purposes.

Personal mail box
Provision of the facility to receive non-urgent (voice and non-voice) messages.

Emergency call
Provision of user facility to transmit an emergency call.

Table 2.2. Traffic information functions.

- Navigation
Provision of information in order to locate a vehicle in the network. Navigation can be either static, using autonomous on-board equipment, or dynamic, using also external infrastructure.

Table 2.2 continued

Autonomous location estimation
Determination of the geographical position of the vehicle with reference to the road network. It can be performed on-board using dead reckoning and map matching. The presence of an on-board map data base enables users to select precise destinations.
Externally assisted location estimation
Determination of the position of a vehicle with the aid of external means such as satellites, cellular mobile radio stations or roadside beacons.
- Optimal route computation
Computes optimal routes on the basis of either individual or collective criteria, and by using either static or dynamic information.
Individual route computation
Computation of an individual route on the basis of criteria selected by the driver, e.g. travel time, travel distance or travel cost. Route optimization can be performed using either static route characteristics or dynamic route and traffic information.
Collective (network) route computation
Provision of a set of optimal routes, by computing optimal paths through the network based on a network control strategy.
- Dynamic route information
Provision of on-line information on road characteristics, traffic and events along a route.
Road works provision of information on type and duration as well as on the expected impact of road works on traffic.
Accidents provision of information on type of accidents with qualifiers such as number of vehicles involved, hazardous materials, length of queue and location etc.
Restrictions provision of information on road or lane closures, contra-flow with narrow lanes etc.
Local warnings provision of direct and timely warnings on dangerous vehicles, abnormal loads, equipment failures, vehicles in wrong direction, obstruction hazards etc.
Traffic conditions provision of information on stationary traffic, length of queue and location, slow traffic etc.
Weather conditions provision of information on weather conditions and forecasts with impacts on road surface, wind conditions, visibility etc.
Environmental conditions provision of direct and timely warnings of adverse and dangerous environmental conditions, including air pollution levels.
Special events provision of information about major events, sports, shows, demonstrations and festivals that affect traffic conditions.
Link travel times provision of information on travel times on road links, based on dynamic traffic information.
- Route guidance
Provision of route recommendations based on individual or collective route computation.
Route presentation
Provision of a displayed description of the optimal route to be followed, which is to be interpreted by the driver.
Route guidance instruction
Provision of a visual or vocal description of the optimal route to be followed with driving recommendations (instructions).

The highlight of all technological innovations in the field of transport telematics is formed by the technological options that provide motorists *during* their trip with all possible forms of assistance for navigation and route selection. The majority of such systems require a (mobile) communication link between vehicle and the external environment. This feature, together with the interface of traffic information systems with the other distinguished functional areas, make the development of traffic information systems a field of primary interest in the development of transport telematics. Following the typology of OECD (1988), these systems are classified into five different categories. These categories differ in the character of *communication links* between the vehicle, road and control centre. These systems are as follows:

Class 1	*Autonomous navigation systems* (no external communication link);
Class 2	*Area broadcast systems* (one-way link from control centre to vehicle);
Class 3	*Local roadside transmitter systems* (one-way link from roadside to vehicle);
Class 4	*Mobile radio systems* (two-way link between control centre and vehicle);
Class 5	*Local roadside transceiver systems* (two-way link between roadside and vehicle).

Besides the type of communication link, the different systems vary also with regard to the type of *interface* with the driver. Basically, there are three types of interfaces: simple directional aids, map display systems and route guidance aids (Jeffery 1990). All of these interfaces require a keyboard for inter alia the inputting of the origin and destination into the system; a microcomputer for reading data inputted by the user and/or data from the external environment, and the subsequent data processing; and finally a display or voice unit.

Simple directional aids show the driver the heading and remaining straight line distance to his destination. The heading information is usually displayed as an arrow. These systems tell the driver something about his actual position in relation to his destination, but are not able to show available or optimal routes.

Map display systems show the driver both the present position and required destination on a continuously updated map display. These map displays are currently based on video display techniques, e.g. liquid crystal displays (LCD) and cathode ray tubes (CRT), while the technology of head-up display (HUD), consisting of information projection on a display in front of the driver, is still in an early stage of development. Map display systems require a high capacity memory unit, such as Compact Disk Read Only Memory (CD-ROM) to contain the network data.

Route guidance aids provide the additional step of identifying and guiding the driver over a kind of optimal route. These require additional data which describe not only the locations of junctions and the course of the roads which make up the network, but also the impedances caused by fluctuating traffic conditions. A suitable algorithm is used to operate on these data with the purpose to deduce an optimal route solution, which can be based on user-dependent criteria, like travel time, distance and travel cost.

The above mentioned five classes of systems will be discussed, in order of increasing complexity of the necessary communication links. This complexity determines the contents and quality of the information presented to the driver.

Autonomous navigation systems

Autonomous navigation systems are essentially self-contained systems for providing drivers navigation assistance, by information on the current location and guidance information. These do not communicate with information sources outside the vehicle. Most of these systems rely on a dead reckoning method to deduce the location. This usually involves fitting the vehicle with both distance and direction sensors, so that the vehicle's progress from a known start position can be continuously monitored and updated.

Area broadcasting

Area broadcasting by radio enables traffic messages to be broadcast over a wide area, typically to about hundred kilometres in diameter. Such systems provide one way communication and can be realized in a number of ways. However, problems related to the capacity of the radio frequency spectrum means that it is unlikely that a suitable frequency may be found for international use. In the past, efforts have been directed at sharing facilities with an existing digital broadcasting system. Possibilities include (one-way) cellular radio systems and satellite broadcasting systems. Cellular radio splits up an area into sub-areas or cells. Traffic information is broadcast on different frequencies in adjacent cells; thus, non-adjacent cells can re-use the same frequencies. Systems in this field can locate vehicles to cells and provide area-specific broadcast information.

The most promising technique for area broadcasts for Europe however, at the present time is the Radio Data System (RDS). This system enables digitally encoded messages to be superimposed on normal VHF-FM radio programme broadcasts. Depending on the in-vehicle software, RDS systems can provide all kinds of service information, weather and road hazard warnings, and act as navigation and route guidance tools.

A considerable measure of agreement has been reached on the data structures and protocols (most likely ALERT-C) for RDS traffic broadcasts in Europe (the system has already been adopted by several major broadcasting companies in Europe), and final standards are set for the messages that will cover the requirements for traffic information broadcasts in all European countries, and the specifications how the messages should be coded. These codes can then be interpreted by the driver by reference to a look-up table of messages stored in the vehicle. Therefore, the system may be internationally operating and the driver can choose to receive messages in his own language.

It is envisaged that RDS traffic broadcasts will also be available to drivers who are not equipped with navigation aids. In this case a cheaper modification will incorporate the look-up table directly into an RDS car radio receiver, so that it can interpret the messages and voice them to the driver as they are broadcast, or put them into a buffer memory for retrieval by the driver at any desired moment. Drivers could then select from the broadcast messages by keying an area code into their receiver, ensuring that they only hear broadcasts about the area in which they are driving. The next generation of car radios will be equipped to decode them.

Local roadside transmitter systems

Local roadside transmitter systems represent a logical extension of the area broadcast systems to provide a higher density of transmitters, ultimately perhaps to equipping individual junctions. Communication technologies used for such systems are mainly microwave and infrared techniques. In this situation they are more versatile than area broadcasting. Three variants exist for this technology. First, roadside transmitters can be used as hazard warning-only transmitters. Second, they can be used as simple location beacons. A major application field of this technology can be found in public transport vehicle location systems, where microwave and infrared transmitters are more advanced options than, for example, coded patterns of buried magnets, short range radio broadcasts and leaky coaxial cables. The third use is as complex beacons to download both hazard warnings and route guidance advice. When a vehicle encounters such a beacon, it turns from 'autonomous mode' into 'guided mode', until it leaves again the guidance network.

Mobile radio systems

Mobile radio systems involve the use of two-way communication links between a control centre and the vehicle. At the moment several mobile radio systems are commercially available. However, these systems are usually limited in terms of the number of vehicles they can serve. Therefore, public radio telephones and, with growing popularity, cellular radio systems and satellite communication systems are already used in several countries.

Satellite communication techniques can provide a location fix on the vehicle and can cover wide areas. The accuracy is not sufficient for route guidance, however.

Cellular radio systems divide coverage areas, i.e. countries or regions, into abutting cells whose radii vary from about 16 km in rural areas down to about 2 km in urban areas. Each cell is then served by a fixed transceiver unit which has available several two-way radio channels over which communication with mobile units can be set up. The cells are generally grouped in clusters of seven which between them share the several hundred channels allocated to the system by the relevant telecommunications authorities. The clusters are then repeated throughout the coverage area. Channels are therefore reused, but not by immediately neighbouring cells. Interference effects are thus minimized, while the use of the available channels is maximized.

The transceivers for each cell are interlinked and computer controlled, so that as soon as a vehicle crosses the boundary from one cell to the next a hand-off procedure occurs automatically. The vehicle unit is then allocated a new channel from those available to the transceiver that serves the next cell - without any apparent break in communication. Moreover, the systems interconnect with the public switched telephone network. Cellular systems therefore effectively extend the public telephone network into vehicles, and anything which can be achieved on the public network can, in principle, be achieved on the cellular system.

The main uses of cellular radio technology may be the downloading of hazard warnings and updating electronic maps. The main difference with the previous technologies however is the extra communication link from the driver to the control centre. This link provides new possibilities. First, on request of the driver, service information and route plans can be provided, and emergency calls can be made. Second, a very important implication of both cellular radio and satellite navigation techniques is that the driver can report back his position and status from time to time. This is a very valuable addition, particularly for operators of vehicle fleets (see Section 2.1.3). Third, these data would also enable some estimates to be made of journey times on the individual roads making up the network, and could therefore prove to be useful for traffic management purposes.

Local roadside transceiver systems

Local roadside transceiver systems use a two-way communication link between roadside units placed on the approaches to major intersections, and special in-vehicle units. The roadside units are controlled by a central computer and respond with a guidance instruction on receipt of a destination code transmitted from a passing vehicle. The guidance instruction is then displayed for the driver, usually on a visual display which mimics the layout of the junction ahead.

All early systems in this field used inductive and buried loop technologies to provide the communication at a precise location along the road. These technologies can act as vehicle detectors, both to interrogate equipped vehicles to learn their trip time between two loops, and to count unequipped vehicles. Loops can therefore be used to collect most of the information needed in order to detect problem areas and to provide dynamic route guidance advice. As in the case of local roadside transmitter systems, recently developed options in this field are based on infrared and microwave communication techniques.

Advanced local roadside transceiver systems can provide in the first place autonomous navigation, using in-vehicle simple directional aids. In a second place, they can act as complex beacons (see the local roadside transmitter systems). Third, these systems have the addition of communicating from the vehicle back to the roadside beacons, so that participating vehicles can communicate their trip times to the infrastructure (without user interference). These data can then be processed by a control centre in order to update the real time guidance advice provided by the roadside units. This feedback feature of these systems has also important implications for the possibilities of traffic monitoring in the field of traffic management.

2.1.2 Public Transport Management

Public transport management deals with the management of public transport services, including fare collection and provision of information to passengers. Telematics assisted functions within this field of transport are summarized in Table 2.3. Strategic planning conducted on the basis of an analysis of operational parameters like demand, revenue, fleet and personnel resources is not included in the range of telematics assisted functions within this area, since it is assumed that this aspect of strategic planning is not an area where clearly telematics applications can directly assist. Indirectly, telematics have an interface with strategic planning, since some of the functions summarized in Table 2.3 provide inputs to strategic planning, e.g. information collection on public transport operations.

Five groups of key technologies can be distinguished in the field of public transport management. Some of these technologies partially overlap with technologies already discussed in the previous sections.

Automatic Vehicle Monitoring (AVM) and Automatic Vehicle Location (AVL) systems

Typical automatic vehicle monitoring and location systems include hard- and software in vehicles and operating centres and communication links between vehicles, roadside and control centre. AVM systems cover a range of functions from vehicle diagnostics and passenger counting, while AVL systems are used for fleet monitoring and control. AVL systems are based on techniques like vehicle odometers, dead reckoning, roadside beacons, radio triangulation and satellite positioning (see the previous section for more details on the necessary specific mobile communication techniques). Two or more of these techniques are often used in tandem in individual applications, since there is no clearly leading technology due to lack of harmonization in the development of AVM and AVL systems.

Interactive passenger information terminals

The application of interactive terminals (and PCs) for public transport information form a substantial part of all systems providing travel information (see Section 2.1.1). Interactive passenger information terminals can be used by travellers or by the staff of enquiry offices and can be located at home, in the office, public places and enquiry offices. The information provided is generally static, i.e. based on current time tables, and can be mainly used for trip planning. Considerable advances have been made in the development of software able to rapidly calculate the optimum route between two points on the public transport network. The use of in home terminals for public transport passenger information is currently confined to the Minitel system (France), which also provides other services such as home shopping.

Table 2.3. Public transport management functions.

- **Scheduling**

Programming all actions for network operation.

Travel time definition

Definition of journey travel times for the construction of timetables, in line with historically recorded data and other operational requirements.

Detailed timetable planning

Planning of all different journeys on all lines and for all operating days, taking also into account demand requirements and cost implications.

Vehicle scheduling

Assignment of vehicles to journeys. Based on defined timetables, vehicle scheduling attempts to optimize the operation by switching vehicles between routes, integrating schedules for a group of routes, and incorporating special journeys.

- **Operations management**

Management of depot and vehicles, crews and incidents and regulating the service.

On-line monitoring

Monitoring of real time location, status and occupancy of vehicle and fleet.

Arrival prediction

Provision of the predicted arrival time of a public transport vehicle at a given point along the route, e.g. bus stop, stop line or traffic light.

Operations control

Preparation and execution of corrective actions based on the on-line vehicle monitoring and standing schedule with the aim of improving service reliability and schedule adherence, either automatically, e.g. automatic request of priority at traffic lights, or via operators.

Assistance to driver

Provision of on-board operational instructions to the driver.

On-demand service provision

Provision of public transport services on request, e.g. taxi services.

User estimation

Provision of the estimate of passenger load (number of passengers) per public transport vehicle.

Table 2.3. Continued

Information on operation
Processing information about the performance of the operation of the network in terms of service (delays, occupancy, disturbance etc.), commercial worth and technical performance of the fleet (technical statistics).

- Passenger information
Supply of useful information to public transport existing and potential users.

Basic information
Provision of information of general nature, independent of a specific passenger trip (e.g. schedules, travel time, network fares, interchanges, P & R information).

Customized trip preparation help
Provision of specific information on request to help a passenger organize a trip. This includes connecting passengers to information centres of other transport modes and reservation/ticketing systems.

Information on current operation
Provision of passengers with actual information on vehicle location, route and destination, waiting times, interchange, delays, service disruptions etc.

- Fare collection
Managing tariffs according to fare policy and requirements of demand management in the overall network.

Operating sales
Issue and delivery of travel documents (single tickets, travel cards and multi-journey tickets).

Validation of travel documents
Validation and verification of passenger entitlement to service.

Fare collection data processing/collection
Transfer and processing of data on ticket sales and validation to provide statistical reports.

- Maintenance
Carrying out routine and non-routine (breakdown) maintenance.

Vehicle diagnostics & data recording
Continuously updating of vehicle logbook (containing vehicle data on technical and performance characteristics, failures, maintenance, age and mileage), providing empirical bases for improvement of vehicle reliability.

Maintenance management
Optimization of maintenance schedules based on relevant criteria (e.g. operational capacity, material reservations and costs), and monitoring their implementation.

Real time passenger information systems

Real time information displays, mainly used to provide real arrival and leaving times of the next series of vehicles, are usually an adjunct to AVM and AVL systems. Such displays can be located at places like public transport stops and interchanges, and in vehicles. The application of real time information in other locations such as in home or via portable terminals is less well developed, but it is likely to become increasingly popular.

Fare collection systems

There have been considerable developments in the last decade in the field of electronic ticket machines, magnetic ticketing and smartcards. The most valid option at present is the use of smartcards. Such cards can be used for integrated payment systems in which the same card is used for public transport and other purposes, e.g. car parking, telephones, etc.

Public transport priority systems

Priority for public transport vehicles in urban areas is becoming increasingly important. Options include public transport priority lanes, gating systems and priority treatment for buses and trams at traffic signals. However, the development of integrated urban traffic control systems which attempt to optimize traffic flows has caused difficulties for public transport priority. Currently, developments are made with the integration of public transport priority measures with urban traffic control techniques.

2.1.3 Freight and Fleet Management

The benefits of information technology innovations in road freight operations (and likewise also in other kinds of fleet operations) result from improved, real-time electronic exchange of, and acting upon, data and information in a wide range of freight management functions or systems. This set of functions is shown in Table 2.4. At the heart of fleet operation lies the effective planning and monitoring of the fleet. Furthermore, management systems exist which deal with dispatch, storage and customs documentation and delivery information. Other systems deal with post-shipment analysis of performance and invoice/payment arrangements. The last area of fleet operations relates to the monitoring, recording and analyzing the vehicle and cargo status, maintenance requirements, provision of warning messages to the driver and emergency call systems.

Table 2.4. Freight and fleet management functions.

- **Logistics and freight management (LM)**

Activities related to the logistics chain from the supplier to the receiver of goods with the associated information and transaction flow. These activities are mainly goods owner concerns.

LM business transactions

Handling of market inquiries, offer and supplier evaluation, just-in-time requirements, contract settlement, invoicing and payments.

Intermodal change planning and preparation

Selection of optimal transport modes and provides necessary booking and other information exchange.

LM operation preparation

Preparation of transport documents, transport orders, customs declarations, dangerous goods declarations and notice of dispatch.

LM operation control

Cargo tracking, shipment and transport status monitoring and delivery information.

LM operation evaluation

Cost and performance follow-up and post-analysis, statistics as feedback to strategic planning.

- **Fleet and resource management (FM)**

All activities directly related to the operational management of the fleet, which are mainly concerns of the fleet operator. Note that the activities listed may not pertain to the management of all types of fleets or may have different names in the management of fleets of different types.

Table 2.4. Continued

FM business transactions

Inquiry acquisition, offer calculation and submission, contract settlement, invoicing and payments.

FM operation planning and preparation

Scheduling of vehicles and drivers, preparation, updating, checking and transferring of documents, task assignment, load planning, operational route planning, third party information (e.g. customs).

FM operation control

Fleet monitoring; special route guidance and traffic information; notification of changes of tasks and route instructions; consignment or passenger pick-up/delivery instructions; trip, cargo vehicle and driver status monitoring; vehicle/cargo tracking/tracing; break-down remedy support; anti-theft tracking of vehicle/cargo. Most of these activities are performed via data communication between vehicle and fixed network stations or home bases, e.g. via satellites.

FM maintenance

Vehicle data compilation and analysis, maintenance planning, material planning, maintenance ordering and evaluation.

FM operation evaluation

Vehicle, fleet and driver staff cost/performance monitoring and follow-up, e.g. vehicle mileage/hours, fuel consumption, repair costs, user fees, driver hours/wages/expenses and pick-up/delivery times.

- Vehicle/cargo management (VM)

Management of individual vehicles, which is mainly the concern of vehicle supervisors and drivers. The activities listed may not pertain to the management of all types of fleets.

VM preparation

Vehicle preparation, trip preparation including weather and traffic information, cargo/storage, equipment and document control.

VM operation

On-board vehicle/cargo/driver status monitoring, vehicle location monitoring, actual route selection, new orders, user fee and other payments, consignment pick-up/delivery and documentation, customs clearance, hazardous goods notification, maintenance actions, general (social) communication and emergency calls.

VM evaluation

Vehicle cost/performance recording, driver hour recording/earnings calculation, deviation and disturbance (traffic, route, vehicle, cargo, delivery/pick-up) reporting.

- Hazardous goods monitoring (HGM)

Monitoring of dangerous goods (chemical, nuclear etc.) and oversize or overweight transports which need special attention.

HGM special route planning

Operational route planning for hazardous goods, which must take special limitations into account depending on the hazard. In critical cases confirmation from responsible authorities may be needed.

HGM communication with traffic control centre

Actual real time reporting of status and location as requested, directly from vehicle to traffic control centre, or indirectly via fleet management centre, and receiving instructions from fleet management or traffic control centre.

HGM special emergency call

Emergency call with high level of priority, depending on the type of hazard.

Information technology innovations in the freight transport sector may assist the above shown management functions by enabling electronic exchange of data and information.

The following four main technologies can be distinguished:
1. electronic passage of ordering and shipping documents among manufacturers or other kinds of shippers, road haulage firms and receivers;
2. automatic identification and status checking of cargo, vehicle and driver;
3. automatic location of vehicles and two-way communications;
4. autonomous navigation.

The associated equipment, systems, services and know-how likely to be used in the transport sector in the medium-term future (5-10 years) are already commercially available or are available as prototypes on the shelves of the transport telematics industries and software houses. In general these include:
- computers and their storage and display media, including on-board equipment;
- specialized computer software;
- automated and semi-automated equipment (computer-controlled inventory systems at the local hub to stock goods to be transported, bar-code readers, voice/data communication terminals for stationary and mobile use, etc.)
- telecommunications infrastructure (telephones, leased lines, satellites and mobile cellular systems) interconnecting the above items and thus permitting one- and/or two-way communications for identification and guidance of the vehicles.

In the following the specific technologies in these application areas will be described in more detail.

Electronic data interchange

The term Electronic Data Interchange (EDI) systems comprises the electronic exchange of transport documents, orders, etc. in a standardized form between road transporters, clients and receivers. Regardless of which networks are used for telecommunications, one long-standing barrier to problem-free electronic data interchange was the lack of generally accepted communications standards. The reason for this situation was that computer manufacturers and data processing equipment suppliers pursue proprietary communication protocols and architectural strategies which make it very difficult, if not impossible, to communicate between information systems of different suppliers.

The basis for open communications between different computer systems is the Open System Interconnection (OSI) reference model of seven layers and conforming protocols. For the user, OSI standards such as those for message handling, file transfer and management, directories and transaction processing from the basis for the exchange of EDI messages. EDI standards are concerned with determining agreed message structures, codes and processes for the transfer of business data, so that there is an exact interpretation of the information being transferred by the participating organizations. EDI links outputs from, and inputs to, applications, across networks between two or more organizations. It therefore requires technical cohesion between participating organizations since different internal processing systems are being logically integrated by the EDI system.

Automatic tracking and tracing

The aim of all automatic identification systems is to make digitized data available to a computer without human processing. Automatic identification (AI) is a component of tracking technologies which are an important part of many advanced communications and information technologies being used to improve the efficiency of freight transportation. Bar codes, magnetic striped cards and smart cards are electronic identification technologies that are already widespread throughout the world. In particular bar codes have already achieved broad acceptance as means of identifying and tracking parcels and other freight, but also other technologies are being developed in this field like character and voice recognition and radio frequency identification (RF-ID). All such systems have the following features:

- a code identifying an object;
- a symbolized representation of this code;
- a support for the symbol (label or tag);
- equipment able to enter the symbol into the support;
- a reader which can make a signal from the support of the symbol;
- a decoder able to decode the signal and to transfer it to a computer.

Advances in radio frequency identification eliminate a major drawback of earlier electronic identification techniques, namely the need for physical contact between the reader and the object being identified. This technology makes it possible to identify objects at highway speeds. The opportunities of these systems are:

- Automatic toll collection without stopping.
- Identification of trucks at plant gates and routing to appropriate loading docks by means of variable message signs. This technology is being employed extensively as part of 'Just-In-Time' (JIT) production concepts being used by automobile assembly plants.
- Identification of intermodal containers and truck trailers at ports, rail heads and trucking terminals, thereby eliminating 'lost' containers and trailers and efficiently matching equipment and loads. For example, an overweight intermodal container requiring a specialized chassis to transport it safely may be identified and matched quickly with the proper trailer to form a legal load.
- Reduction of delays at regulatory stopping points such as border crossings and inspection points by the identification of trucks in advance. Once identified, the vehicle's appliance with applicable regulations (payment of registration fees, fuel tax permits, oversize or overweight permits, etc.) may be checked.
- Identification of illegal drivers. Radio frequency identification technologies, combined with smart card driver's licenses and electronic on-board recording devices, may eliminate the need to stop vehicles in order to monitor driver's compliancies with hours-of-service and licensing regulations.

Automatic vehicle location and two-way communication systems

Automatic vehicle location (AVL) systems make it possible for logistic operators to monitor movements of vehicles. Two-way communication systems establish a bi-directional communication between the operator and the drivers to very long distances. Some existing products offer both automatic vehicle location as well as the possibility for satellite communication. AVL systems and satellite communication systems envisage software equipment both in vehicles and in the fleet operator centre for tracing purposes. The main features of these technologies have been treated in previous sections.

Basically, the range of available technical options in the field of mobile transmission systems encompasses at the one hand technologies based on public networks and at the other hand private mobile radio systems. Technologies based on public networks comprise mobile telephony, satellite communication systems and systems using cellular radio networks. Private mobile radio systems comprise - in addition to private radio - also the new radio technologies of trunking systems (which is quite similar to cellular radio technique) and paging systems. Although the widely-used private radios are reliable and cost-effective technologies for trucks, these systems may have only a limited range. There is therefore a clear need to extend the range of these systems to areas which have a size of 150 km or more. This requirement is met by these new technologies. An additional primary feature of cellular radio technique is that it can serve large numbers of users while the frequency spectrum available for radio communications is restricted.

The limited geographical range has been until recently also a shortcoming of mobile telephony, because of its dependence on local telecommunication infrastructure. It is important to note however, that the introduction of digital mobile telephony systems, and in particular those systems based on the so-called GSM technology (Global Systems for Mobile telecommunications) have largely the same user possibilities as satellite communication. GSM is a digital system, but it is moreover adopted in an important number of European countries, more or less functioning as a European standard. Hence, mobile telephony is acquiring the same advantages as satellite communication - digital technology and European wide use-, while this system is relatively much cheaper for the user than satellite communication applications.

Navigation and route guidance systems

Delay in traffic is costly to the freight sector. Trucks in particular are disproportionately affected by traffic congestion in relation to private vehicles. The monetary advantages of dynamic traffic information by truck drivers may be higher than in the case of private users, since in logistic operations there are also shadow costs of delays in deliveries and empty load space.

Navigation and route guidance systems are telematics applications that assist drivers on their trip with their navigation task, by providing data on current location and traffic conditions. Basically, the systems designed for private vehicles also apply to freight vehicles. An important feature of the freight sector however is its geographical orientation which is wider than average commuting distances. Therefore, truck drivers require navigation support systems which are able to cover relatively large surfaces. In addition to road-vehicle communication systems which focus on urban areas, there is thus considerable scope for long-distance terrestrial radio navigation systems and satellite navigation systems which are specifically developed for truck operations.

2.1.4 Road and Traffic Management

The management of roads and traffic is a task of national and regional responsible governmental bodies and infrastructure authorities and operators. It involves a wide series of relevant applications. In short, these areas involve the on-line monitoring of road and traffic conditions, the gathering of data, the subsequent processing of these data, and traffic control measures, which are based on monitoring data (including forecasts) (Klijnhout 1990). In Tables 2.5 and 2.6 a more detailed overview is given of the relevant functions within the field of management of roads and traffic.

Table 2.5. Road management functions.

- Ambient conditions monitoring
Monitors and provides weather and other ambient condition data relevant for road transport; the data are filtered for spatial relevance.

Meteo monitoring
Monitoring and filtering variables like temperature, relative humidity, wind and precipitation with respect to relevance for road conditions.

Visibility monitoring
Local roadside monitoring of visibility.

Area pollution monitoring
Monitoring of pollution conditions based on local sensors.

- Road status monitoring
Monitors and provides data regarding road surface conditions and status of roadside equipment.

Friction monitoring
Monitoring of friction data, which are calculated from meteo data and refined by data from road-based and vehicle-based friction sensors.

Infrastructure monitoring
Monitoring of the status of infrastructure, e.g. roads, bridges etc.

Road infrastructure capacity reduction / road works effect calculation
Monitoring and provision of data on calculated effects of road capacity reductions due to planned maintenance activities.

Weigh-in-motion monitoring
Monitoring and collection of data from weigh-in-motion stations.

- Forecasting
Provision of forecasts of variables related to pollution, special events and ice forming.

Pollution forecasting
Forecasting and provision of information on future pollution situations.

Special events forecasting
Forecasting of location, duration, expected traffic volume and other relevant characteristics of scheduled special events, e.g. sport events, concerts, strikes, demonstrations, special transports such as long and heavy loads, military convoys etc.

Early ice warning
Detection of and warning for situations that may lead to forming of black ice on the road, on the basis of data on ambient and road surface conditions.

- Rescue service and maintenance management
Providing of continuous surveillance of needs for rescue services and maintenance, and the initiation and management of appropriate action in response to these.

Emergency call management
Collection and processing emergency calls in order to provide appropriate schedules for intervention.

Hazardous goods location monitoring
Receiving location reports and maintaining overview of hazardous goods transport locations within responsibility areas.

Hazardous goods rescue service
Operating and managing the required rescue service based on type of hazardous good.

Maintenance management
Operating and managing the required short term road maintenance activities.

Table 2.6. Traffic management functions.

- Road section traffic control
All measures designed to control traffic flows on road sections.

Section state monitoring
Generation of data on traffic characteristics, e.g. average speed and density; pre-processing data for flow estimation; monitoring congestion.

Incident/accident detection and identification
Detection and identification of abnormal occurrences relevant for the state of traffic and travel.

Section control computation
Computation of control measures like speed recommendations and limitations, lane recommendation and allocation and ramp metering for a road section on the basis of control strategies.

Section control actuating
Activation of control devices, either individual or public, e.g. traffic lights and variable message signs.

Local speed enforcement
Control measures on road sections of high potential risk, e.g. in residential areas or near schools and at dangerous intersections.

- Road junction traffic control
All control measures that are specific to road intersections, in both urban and inter-urban areas.

Junction state monitoring
Monitoring of traffic status at junctions, including monitoring of vulnerable road users.

Junction control computation
Effectuation of control by the integration of empirical data and model-based estimations. Measures include modification of priorities in control actions, i.e. to grant priority to selected vehicles such as public transport, police, rescue and fire brigade, or optimization for pedestrians in specific areas.

Junction control actuating
Operation of control devices, such as actuating of traffic lights and variable message signs.

- Network traffic control
Application of control measures to road sections and junctions in an area wide network, e.g. cities or motorway networks.

Network state surveillance
Monitoring and compiling the actual traffic load of the network (flows, journey times, etc.) and updating of the network status database.

Origin/destination computation and route assignment estimation
Computation of O/D paths through a network, taking into account constraints imposed by traffic control strategies and actual occurrences, such as traffic congestion and incidents. These computations are repeated frequently in order for control measures (which can range from route advice to individual route guidance) to be based on an up-to-date status of traffic.

Demand prediction
Prediction of traffic load on a network based on historical (statistical) data and current information from adjacent networks.

Traffic prediction
Prediction of trends based on the current state, O/D computation and special events forecasting.

Network control computation
Assessment of control plans at the network level, with the specific aim of giving coordination rules to other functions like section and junction control.

Table 2.6. continued

- Localized area traffic control
Application of measures to specific locations or areas in the network, e.g. tunnels, which require dedicated strategies due to their specific operational environment.

Tidal flow control
Control of the reservation of lanes for exclusive use in one direction at one time and the reservation of the same lanes for the reverse direction at another time, in order to accommodate the alternating flow of traffic (reverse lane control).

Ramp control
Implementation of control measures for merging traffic flows at motorway junctions, taking into account traffic demand and the actual capacity of the road section ahead. The measures are based on calculations from traffic monitoring data, such as ramp metering.

Tunnel traffic control
Implementation of control measures that meet the specific requirements of "tunnel environments", i.e. higher update frequencies in monitoring and enhanced signalling.

Bridge traffic control
As for tunnels, with possible additional wind warning.

Lane management
Implementation of control measures reserving certain traffic lanes exclusively to specific classes of vehicles, e.g. high occupancy vehicles and buses.

- Policing/enforcing
Provision of measures performed by enforcement authorities to achieve compliance with (road) transport laws, rules and regulations.

Automatic vehicle identification
Provision of identification of vehicles, e.g. via electronic license plates.

Violation registration
Provision of evidence of traffic offenses, e.g excess speeding, overload, illegal overtaking, violating red lights, use of restricted lanes and parking violation. Also performs registration of corresponding penalties.

The different functions in this field require extensive data collection, transmission and processing techniques for the support of management decisions and to provide real time information to drivers. Examples are traffic sensors for automatic data collection (counts, speed, occupancy), fog and ice sensors, tunnel and bridge control systems, Variable Message Signs (VMS) and road-side emergency call posts. They can be used directly for decision support by automatic action or via managers. Ideally, all such systems should be integrated with each other into a so-called Integrated Road Transport Environment (IRTE).

Monitoring systems

Monitoring systems comprise inductive loop detectors (buried in the carriageway), the successors to pneumatic tubes which are still being used, and Closed Circuit Television (CCTV) systems based on video sensors, which are systems being used for real time data collection on traffic counts, congestion and incident detection on roads. At present inductive loops are the most widespread monitoring systems. Video sensors are now being advanced by a new generation of image processing cameras. In conjunction with conventional reporting such as patrols and telephone calls, these devices are proving very effective in the instant recording of congestion and incidents in zones of up to a few hundred metres. The most advanced variant of this technology, ultrasonic sensors, are in operation only in Japan. In the longer term, it is envisaged that these basic data needed for traffic management will be collected by means of communication between vehicles and traffic control centres (such communication options are in detail discussed in Section 2.1.1).

Variable Message Signs (VMS)

There has been a recent expansion in the use of roadside Variable Message Signs, of which more than 200 systems already exist in Europe. They are used to provide car drivers with various kinds of dynamic information, like information on road works, incidents, lane restrictions, weather, environmental and traffic conditions. Several different display techniques are being used, ranging from the older mechanical devices to the more advanced LCD and light emitting diode (LED) technologies. The information presented on these signs vary from short messages (signalling, e.g. advisory speeds) to more complex messages (e.g. indications of congestion ahead and route advice). Systems that are connected with monitoring systems can provide the most accurate and reliable information.

Ramp metering

Ramp metering is the merging of a traffic flow onto a congested motorway by means of continuous one-by-one access, controlled by a traffic light. Merging is smoothed so that the onset of congestion at the respective strategic road junction is delayed. The available capacity is closely matched with demand to ensure that the traffic on the motorway continues to flow.

Emergency Telephones

On most motorways, a network of emergency telephones is available. These telephones serve to help motorists in cases of incidents or technical problems. They also help to identify where potential disruptions may occur and the police are informed.

Priority Lanes

Priority lanes are systems that enable specific road user groups to access part of the strategic road network with some form of priority of particular interest to policy makers. Possibilities include the priority treatment to buses at traffic lights or bus lanes on congested (motor)ways in order to stimulate public transport. Travel time is then reduced and a competitive advantage is realized. Other examples are High Occupancy Vehicle (HOV) lanes (or carpool lanes), i.e. special lanes with access only for cars occupied by more than one person, and Heavy Goods Vehicles (HGV) lanes.

In summary, there are many telematics products and services on the market for road and traffic control. They are being continually improved to make them more dynamic in operation and with increasing functional integration of databases. The software supply industry attached to this integration is substantial.

2.1.5 Demand Management

It must be realized that the various forms of operational traffic control will never be able to cope with traffic when demand significantly exceeds capacity. In such cases of oversaturation, measures to restrict traffic demand will be a rigorous but valid option to alleviate traffic congestion without an extension of the existing infrastructure. In the future matching of traffic demand with the available infrastructure, telematics can provide new means of managing traffic demand by various functions (Table 2.7). They are chiefly restrictive in terms of limiting traffic access at particular times of the day, or by using pricing measures as a restraint on car use or access to motorway segments and urban zones. In the longer term, demand management has clear links to land use and facility planning which should seek to minimize trip lengths and dependence on the use of the car. Close links should be established between demand management functions as described in Table 2.7 and other functional areas in order to constitute integrated transport strategies. Examples may be parking policies and the supply of travel and traffic information that meet demand management objectives.

Table 2.7. Demand management functions.

- Demand restraints

Measures to reduce congestion in sensitive or overloaded areas and to control modal split.

Area access restriction

Medium-term measures to protect areas limited in capacity from unwanted traffic, e.g. by a 'traffic collar', banning or limiting entry.

Route diversions

Medium and long-term planning of diversion of traffic to alternative routes, also with respect to heavy vehicles and hazardous goods vehicles.

Road pricing

General term for different measures of charging for road use derived from policy decisions, e.g. taxation schemes, urban congestion pricing, urban cordon pricing, peak-hour pricing, inter-urban road tolling, bridge and tunnel tolling etc.

Area parking strategy

Measures to implement operational parking strategies for specific areas, e.g. preferences for residents, time and period limitations.

- Supply control

Supply control includes all medium and long term measures to reduce congestion via the provision of alternative travel means.

Car pooling

The facilitation of the sharing of one car by several persons (on a shared cost basis) by the provision of information and possible incentives. This is a form of paratransit organized for a group of persons having similar origin/destination combinations.

Modal interchange

Provision means for the integrated use of different transport modes for one trip, e.g. information, booking mechanisms and parking, especially changing between private car and public transport, such as Park & Ride and multi-modal freight transport.

The technological developments of telematics technologies for road pricing and congestion control are moving faster than the political and public acceptance of such measures as a strategy for traffic demand management. In contrast, debiting in parking or public transport applications is likely to have greater acceptability. Much research and development has been undertaken into the relevant technologies. Despite this, few cities are undertaking trials of telematics applications assisting these demand management measures, although an increasing number are introducing traffic restrictions in urban centres.

The most important technological developments in the area of demand management are Automatic Vehicle Identification (AVI) and automatic debiting technologies which can be applied to several of the functions described in Table 2.7, and in particular to area access restrictions and road pricing. These technologies are now coming on the market. They use AVI tags, which are vehicle-bound devices which can communicate with signal transmitters on the road-side. They can provide access for certain classes of vehicles or user groups in restricted areas (e.g. residents, special needs groups and drivers of 'clean vehicles'), fast automatic toll collection and traffic metering in a given area. First generation electronic AVI tags are read by a ground antenna and compared with a file; authorized vehicles are then allowed to pass. Such tags can be used for regular users of a toll road and require low speeds. Second generation AVI tags are under development based on two-way short range (up to a few hundred meters), high capacity microwave communication links, which has become the most important communication method for AVI technologies (in addition to microwaves, infrared beacons are used, but mainly for the identification of freight vehicles). Such facilities would apply to closed toll systems for payment by deduction, or congestion meters which could incorporate a smart card reader. Third generation equipment is currently under trial, consisting of multi-service, multi-use and multi-network devices for various applications.

2.1.6 Parking Management

Parking management is essentially an element from the fields of demand management, traffic control and travel information. Telematics has a large potential to improve the efficiency of parking systems in urban areas through real time monitoring of car park volumes, real time forecasting of available parking space and the dissemination of information to drivers in advance on available parking space and alternative locations such as Park & Ride (P&R) sites. Within car parks, systems are available to guide vehicles to an empty space to avoid cruising vehicles within the parking area. Additional telematics functions in the field of parking management are related to reservation and debiting of parking charges and enforcement/penalty systems for both on and off street parking spaces (Table 2.8).

The particular telematics technologies used for these functions are similar to those applied in other areas of demand management, traffic monitoring and traffic and travel information.

Table 2.8. Parking management functions.

- Parking space management
Monitoring and providing data on parking facilities.

Parking entry/exit monitoring
On-line monitoring of the actual state of occupancy of parking facilities.

Parking occupancy prediction
Provision of forecasts of parking occupancy, based on factors like the actual traffic state, special shopping days and events.

Parking availability control
Implementation of measures to control area-wide parking facilities according to principles of demand management.

- Parking guidance
Wide- and local-area guidance of drivers of cars, trucks and buses to parking facilities (including Park & Ride). This may be based on variable message signs as well as on radio communication systems.

Dynamic parking information
Provision of real time information on parking space availability to drivers.

Network parking guidance
Provision of wide-area guidance information related to parking facilities with available parking space.

Final destination guidance
Guidance of drivers within a parking facility to the parking place assigned to the vehicle.

- Parking reservation and payment
Management of parking areas, commercial or non-commercial.

Parking space booking
Provision of parking reservation in advance, both pre-trip and on-trip.

Parking payment
Provision of means for charging parking fees, e.g. magnetic cards and automatic debiting systems.

2.1.7 Driver Assistance and Cooperative Driving

The functional area of driver assistance and cooperative driving embraces the monitoring of drivers, vehicles and surroundings, and provides direct or indirect assistance for driving, either by warning stimuli to the driver or by partial or full automatic control of the vehicle. This control of the driving task is generally regarded as a long-term option. It is envisaged that vehicle control in its ultimate form, i.e. without driver interference in the driving task, implies a complete new architecture of vehicles and infrastructure resulting in the end more or less in a new transport mode. At present the enabling technologies for vehicle control are still in early development phases. The functions that these will be able to serve are summarized in Table 2.9.

The systems that are already commercially available in these fields include electronic devices such as anti-lock brakes, speed control and active suspension, which are available as either standard or optional equipment on both private cars and commercial vehicles. Products that combine various functions are at present in general in an early stage of development. The key technologies for driver assistance and cooperative driving are now discussed.

Automotive radar systems

Radar technology is the main option for the continuous monitoring of the distance and speed between vehicles on the road. Such radar systems consist of a transceiver fixed on the vehicle that continuously transmits signals which reflects off vehicles ahead. Communication means are potentially low power microwave signals and millimetre-wave signals. The return signal is processed and combined with information about the host vehicle, such as speed and direction. Sophisticated algorithms then process the data and determine if hazardous situations exist. If there is danger of collision, the driver can be warned by visual and/or audible signals to take evasive action. These radar systems can also be combined with speed control systems, by which a safe distance from the traffic ahead is contained (intelligent cruise control), and with (standard) Anti-lock Braking Systems (ABS), by which in potentially hazardous situations the braking process is shortened.

Table 2.9. Driver assistance and cooperative driving functions.

- Monitoring environment & road

Acquisition of information on the current status of the immediate environment of the vehicle.

Road surface and marking monitoring

Monitoring of road surface conditions as well as road lane markings (no combination with meteo data is intended).

Road geometry monitoring

Monitoring the spatial dimension of the zone ahead of the vehicle to establish manoeuvring restrictions.

Visibility monitoring

Monitoring of reduced visibility caused by adverse weather conditions such as fog, rain or snowfall.

Road regulations monitoring

Monitoring of roadside information such as traffic lights and signals.

- Monitoring driver

Observation and evaluation of the driver's control of the vehicle, his psychological condition and possible deviations from normal behaviour.

Driver status monitoring

Monitoring and evaluation of driver's handling of vehicle control devices, his psychological condition and deviations from normal behaviour.

Creation of driver profile & identification of trends

Producing statistics on driver's performance and identification of trends.

- Monitoring vehicle

Acquisition and processing data on vehicle dynamics and operational status to diagnose and predict vehicle failures and vehicle dynamic behaviour.

Vehicle dynamics monitoring

Monitoring of vehicle dynamic status, e.g. speed.

Vehicle operational status monitoring

Monitoring operational status of the vehicle, e.g. identification of current vehicle parameters, degradation or failure.

- Vision enhancement

Improvement of the visibility of the driving scene by autonomous and non-cooperative means, in sub-normal visibility conditions by providing direct visual information to the driver.

Table 2.9 Continued

- Collision risk estimation
Detection of potential obstacles in relation to the dynamics and predicted trajectory of a moving vehicle for collision avoidance.

Relative position determination
Measurement of the relative positions of neighbouring vehicles.

Conflict zone monitoring & trajectory prediction
Detection and monitoring of the dynamic status of other road users within the manoeuvring zone of the vehicle in order to identify trajectories that may lead to conflicts. This includes detection of 'obstacles', e.g. vulnerable road users and static objects.

Safety margin determination
Continuous determination of the range of vehicle performance within which stable driving manoeuvres can be performed, considering all relevant influencing factors related to the vehicle and the environment, both in actual and predictive manner. The steps involved are determination of maximum potential limits; determination of the actual and predicted position of the driving situation; and determination of safety margins.

Critical course determination
Continuous determination of a safe trajectory with regards to road boundaries, stationary and moving objects. This consists of determination of own trajectory; identification of potential collisions with obstacles; and identification of a course which minimizes the risk of collision.

- Dynamic vehicle control
Control of lateral, longitudinal and vertical dynamic behaviour of the vehicle to influence the vehicle in order to stabilize driving according to the demand of the driver or higher-level control system, with priority given to safety requirements. Actuators are controlled in accordance with the control strategy for the actual situation; this ranges from advice to the driver on supportive actions to fully automated control.

Lateral activation/control
Dynamic guidance of lateral movements of the vehicle with regards to lane or road side in order to maintain the appropriate trajectory.

Longitudinal activation/control
Dynamic speed and distance control (cruise control).

- Dialogue management
Dynamic optimization of the interface between driver and machine regarding priority and sequence of information presented to the driver through any type of interface.

User presentation
Identification of priorities of incoming information, optimization and integration of information for actual user interface, e.g. less important situational information can be delayed in order to decrease the driver's workload at any point in time.

Driver tutoring
Provision of feedback to the driver regarding non-compliance with traffic rules, regulations and a safe dynamic driving status.

Driver status monitoring systems

Driver status monitoring systems are designed to detect deleterious changes in driver behaviour through monitoring continuously the range and frequency of deviations from the driver normal performances. The driver status can be monitored on the basis of driver parameters and vehicle parameters. Driver parameters can be derived from physiological measurements by means of electro-encephalograms, electro-oculograms and video camera views of the driver's face. Vehicle parameters can be recorded, such as the steering wheel angle, speed, acceleration (longitudinal, lateral and vertical), together with the force on the accelerator pedal and the car temperature. Combined with data from an headway monitoring system, these data can be interpreted and analyzed by a neural network system. Such a neural network system is capable of updating an adaptive database which contains normative rules concerning driving behaviour with local information, and of comparing registered driver behaviour with these rules. Deviations are converted into messages feedbacked to the driver.

History and instruction module systems

History and instruction module systems aim to improve driver's behaviour by providing the driver with information about how well manoeuvres have been performed, in a way which takes into account the experience and past performance of the driver. These systems comprise on-board instructional support systems which make use of data routinely collected by other driver assistance systems, e.g. radar detection and driver status monitoring. The system processes this information, determines whether the driver requires some feedback on his performance, what the nature of this feedback should be, and how it should be presented.

Dialogue management systems

Dialogue management systems comprise enhanced man-machine interfaces, which manage and supervise all communication between the driver and various telematics systems. By means of dialogue control algorithms and workload estimators these systems can adapt the presentation of messages according to the status of the driver's workload.

2.2 Research Initiatives in the Field of Transport Telematics

In this section the aims and contents of some major telematics research programmes will be discussed in leading countries in the area of transport telecommunications: Europe, the United States and Japan.

2.2.1 Europe

PROMETHEUS

PROMETHEUS (Programme for European Traffic with Highest Efficiency and Unprecedented Safety) has been a seven year research project (1987-1994) within the European EUREKA (European Research Coordination Agency) programme. PROMETHEUS was principally a private sector initiative aimed at developing a uniform European traffic system incorporating transport telematics technologies. In the programme participated technology suppliers, university research institutions and a consortium of European automobile companies that included amongst others Ford, Jaguar, Rolls Royce, Renault, Volvo, Saab, Daimler Benz, Volkswagen, BMW, and

Fiat. The programme had a steering committee consisting of eleven automobile companies.

The main objective of this research was to make road transport significantly safer. In the programme 'intelligent vehicles' and electronic traffic flow detectors were designed that improve communication between drivers. Rather than taking individual elements of road transport, PROMETHEUS considered the system as a whole, establishing close relationships between drivers, vehicles and roads. PROMETHEUS involved a number of field trials of technologies in the functional transport telematics area of driver assistance and cooperative driving. The overall project was broken down into three areas of industrial research and four areas of basic research. The industrial research areas were Pro-car (use of on-board, self-contained technologies to monitor vehicle performance and assist the driver), Pro-net (communication between vehicles) and Pro-road (communication between vehicle and road). The four basic research areas were Pro-art (use of artificial intelligence), Pro-chip (use of micro-electronic components for the various sub-systems), Pro-com (communication between vehicle and driver, other vehicles and roads) and Pro-gen (evaluation of the general impact on the traffic environment).

The expected beneficial impacts of the PROMETHEUS programme were to reduce European traffic casualties by 50% by the year 2000, an improvement in transport efficiency of 20% and a 15% decrease in motorized transport within a city.

DRIVE

DRIVE (Dedicated Road Infrastructure for Vehicle Safety in Europe) is a research programme within the European STIG (Systèmes Télématiques d'Intérêt Général) programme, started in 1989. The three overall objectives of DRIVE are the improvements of transport efficiency and safety as well as the environment (EC 1993). The focus of research in this initiative has until now been wider than the parallel programmes in America and Japan. The research carried out in the DRIVE programme has included aspects of mode and destination choices, public transport users and pedestrians, while in the United States and Japan the focus has been restricted to the effects of information on route and departure time choices.

The DRIVE I programme (1989-1991) under the Second Framework Programme of the European Community focused on a common European road transport environment in which drivers are better informed and where intelligent vehicles interact with the infrastructure. Participants were from the research and development sector (both private and governmental), and from the industry. Projects were arranged into four main groups respectively in charge of general approach and modelling, behavioural aspects and traffic safety, traffic control and services/telecommunications/databases. In addition, efforts were carried out by the project SECFO (Systems Engineering and Consensus Formation Office) to obtain consensus at the strategic level, involving also actors who were not directly involved in DRIVE but were nevertheless affected by the development of transport telematics systems. The collaboration of participants in DRIVE is regarded as having been successful, both in its technical results as well as the contribution made to reducing uncertainties of moving towards transport telematics implementation. This programme led to a number of prototypes and concepts, including the SOCRATES cellular radio based route guidance system, the ALERT-C prestandard for RDS-TMC protocol, smart cards for automatic debiting for moving vehicles and a special frequency allocation for traffic applications.

The development of technologies with an exploratory focus in DRIVE I was followed up by their validation, assessment and integration during 1992-1995 in the DRIVE II programme under the Third Framework Programme of the European Commission. Projects in this programme focused on preparing the implementation of transport telematics, expected to lead towards an Integrated Road Transport Environment (IRTE), while the range of participants was broadened to include a significant number of city and regional authorities. The three main objectives of DRIVE I continued in DRIVE II. The framework of the research carried out in this programme was embodied in seven transport telematics areas of major operational interest, parallel to the classification used in Section 2.1. A number of core activities were carried out among which an important role was played by CORD (Strategic Assessment of Advanced Transport Telematics Implementation). This project assisted in the establishment of consensus between projects on common functional specifications and common practices on the strategies to be used by the pilot projects. This work was intended to lead towards the specification of a system architecture for transport telematics which would be acceptable throughout Europe.

Table 2.10. Frequencies of inter-urban road telematics applications tested in the European DRIVE II programme (1992-1995).

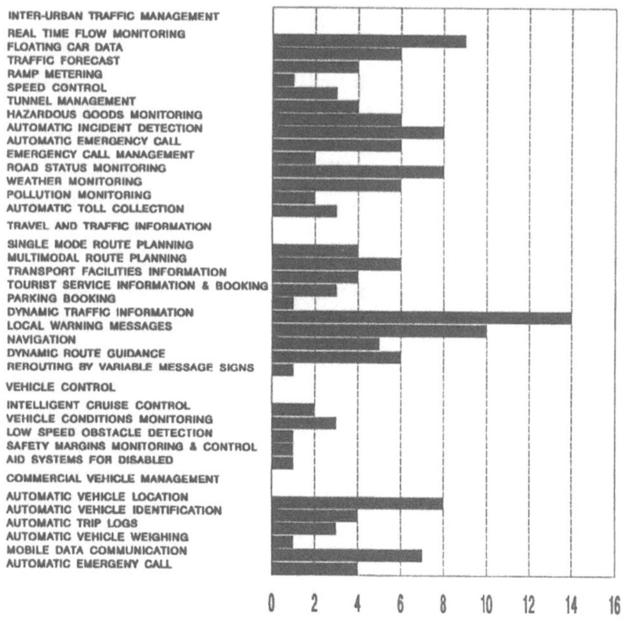

Source: Filippi et al. (1994).

An idea of the relative effort put into the development of the various technologies can be derived from Table 2.10, in which the number of pilot projects are shown dealing with transport telematics applications in the inter-urban field. There was a clear focus on technologies providing dynamic traffic information and local warning messages. This was followed by applications collecting the necessary source information for the provision of dynamic information on traffic and road conditions.

The achievements of DRIVE I and II have helped to promote collaboration and bring order to the diversity of parallel developments in transport telematics in Europe.

In the new Fourth Framework Programme of the European Community, the plan of the DRIVE III programme is to take the process further by promoting essential R&D and more substantial validation trials including market perspectives. Furthermore, the scope will be extended to air, rail and waterborne transport. The emphasis will be inter alia on multimodal services delivered to transport users (private and commercial). The programme will also cover the validation and assessment of integrated transport telematics services where suitably sized pilots can be expected to investigate the user-related issues for which a feedback to technological research is required. The response of the end-user in terms of travel behaviour and in terms of the potential market will thus have a stronger focus in this future research.

2.2.2 United States

Research on transport telematics in the United States began over two decades ago (MARTA 1995). Electronic route guidance systems and automated highways were areas of early research in the 1960s. After a substantial reduction of the federal government's role in civilian technology development during the early 1980s came a revival of research activities in the late 1980s, parallel to the large transport telematics research programmes in Europe and Japan. Since 1987 the term IVHS (Intelligent Vehicle-Highway Systems) has been in use. The renewed movement in IVHS was initiated by an informal group known as Mobility 2000. The activities of this group led to the initiative for a non-profit educational and scientific association called IVHS America (Intelligent Vehicle-Highway Society of America) in 1990.

Two important governmental organizations are the operating administrations of the Department of Transportation (DOT), the Federal Highway administration (FHWA) and the National Highway Traffic Safety Administration (NHTSA). These carry out research and development activities related to IVHS, each in selected areas. While the national IVHS strategic plan has provided a solid foundation for transport telematics research in the United States, a more specific Programme Plan is still under the cooperative development of the public and private sectors. The Programme Plan describes the specific steps to be taken through the stages of R&D, field tests, and deployment guidelines for the IVHS user services, and to integrate them in an open national IVHS system architecture to reach the programme goals of increasing travel efficiency, safety and environmental quality (IVHS America 1992).

The classification of main categories of transport telematics applications in IVHS is only slightly different from the one used in DRIVE: ATMS (Advanced Traffic Management Systems), ADIS (Advanced Driver Information Systems), FFCO (Freight and Fleet Control Operations) and AVCS (Automated Vehicle Control Systems).

The political reality in the United States has been that practically all authorized federal support for IVHS has been earmarked by Congress to conduct these tests on specific sites. Among the more than 40 major field tests are TravTek in Orlando which involves 100 vehicles in dynamic route guidance and motorist information, ADVANCE in suburban Chicago which includes 5,000 vehicles to serve as traffic probes, and the HELP/Crescent project along the Pacific coast and the Southwest US to permit non-stop commercial vehicle operations. In the future, most of the authorized federal funds will be used to support field tests in the 'priority corridors' in Northeast US, Chicago area, Houston area and Southern California.

Since the major resources for IVHS deployment are supposed to draw from non-federal sources, early deployment of IVHS does not need the completion of the national programme plan. The main products of the IVHS activities until now have been in the area of Automatic Vehicle Identification (AVI), mostly for toll collection, and Automatic Vehicle Location (AVL) (mostly for commercial vehicle operations). Low-cost options for IVHS services, including the Radio Broadcast Data System (RBDS) and hand-held traveller information systems, are expected to be pushed by the private sector in the near future.

2.2.3 Japan

Japan is the country where until now most impressive progress has been made in the development and implementation of transport telematics technologies. This is largely caused by the extremely congested traffic situation. The combination of a limited road network and densely populated areas in this country has resulted in heavy congestion throughout most of the day: the average journey speed during peak periods on important roads in urban environments is only around 20 km/h (Uetakaya et al. 1992). The country's decision to embark on a programme for introducing information and communication technologies to transportation was taken already in the 1950s. The Comprehensive Automobile Traffic Control System (CACS) has been the first big challenge in Japan, and probably in the world, to implement advanced information devices for vehicles and roadside facilities. The objective of this project, which was running from 1973 to 1979, was to contribute to the improvement of automobile traffic by developing technologies such as dynamic route guidance, hazard and warning displays etc. This was followed by another programme, the Automobile Traffic Information and Control System (1978-1985), which culminated in a demonstration to bus passengers of in-vehicle information systems using ground to vehicle data communications.

Up to 1985 there has been heavy government funding. The private industry began to contribute significantly through the projects RACS (Road/Automobile Communication System) and AMTICS (Advanced Mobile Traffic Information and Communication System), which have been two projects focusing on the same transport telematics functions but with different communication technologies (Kawashima 1990 and 1992). RACS developed a beacon-based communication system and AMTICS developed a mobile data communication system. The projects run from 1986 and 1987, respectively. In 1991 another large research project was launched named Vehicle Information and Communication Systems (VICS), a grouping of over 200 Japanese companies, which aimed to develop advanced information systems for travellers. This project was an extension of and combined the outputs of the earlier projects RACS and AMTICS.

A similar typology of transport telematics technologies is used as the European one. The main distinguished areas of technologies are ATMS (Advanced Traffic Management Systems), ATIS (Advanced Traveler Information Systems), AVCS (Advanced Vehicle Control Systems) and CVO (Commercial Vehicle Operations).

Various transport telematics applications developed in these research programmes have already been implemented on a large scale in early stages. Examples are the computerized police traffic control centres in the 74 major Japanese cities (since 1985), automated variable message signs (since 1973), roadside radio (since 1983), and the recently introduced Highway Advisory Radio (HAR) for urban motorways. Furthermore, the semi-privatized expressway corporations benefit from automated tolls revenue. The Japanese are also well advanced in the development of driver assistance and cooperative driving systems. A highlighting example is the development of the Personal Vehicle System (PVS), a fully automated way of driving (Kurami et al. 1991).

In Japan there are many governmental agencies, major coordinating bodies and leading private firms involved in developing transport telematics technologies, which means that projects tend to develop in a competitive environment. Each ministry uses a research association that will act as its agent and project coordinator. The result of these many actors in this single country is a complex process of coordination and liaison between the many transport telematics related programmes.

The governmental bodies include the Ministries of Post and Telecommunications (MPT), Transport (MOT) and International Trade and Industry (MITI) and the National Policy Agency (NPA). Major coordinating bodies include the Association of Electronic Technology for Automobile Traffic and Driving (JSK), the Highway Industry Development Organization (HIDO), the Japan Traffic Management Technology Association (JTMTA) and the Japan Digital Road Map Association (JDRMA). The strong involvement of the automobile industry is reflected by the transport telematics research activities of Nissan, Toyota, Honda, Mazda and Mitsubishi, who have each equipped new cars with telematics functions and have brought these into the market.

Major future research initiatives are the Advanced Road Traffic Systems (ARTS) and the Super Smart Vehicle System (SSVS) programmes. ARTS will embrace inter alia projects on automatic toll collection, road alignment information, road surface detection, surrounding object detection and vehicle headway control systems. This scheme has a 10 to 20 year time horizon. The Japanese view on the future of the urban road traffic environment is embodied in the concept of the SSVS. SSVS describes a scenario in which all vehicles act as a cooperative system, harmonized with people and their society (Association of Electronic Technology for Automobile Traffic and Driving 1992). The scenario includes interactive liaison between inter-vehicle functions and driver-to-vehicle functions and also should ensure close cooperation between vehicles and road functions. It is anticipated that the scenario envisaged will be realized within thirty years from now.

2.3 Technological Development Prospects

At the present time the rate of technological development of transport telematics varies largely across the different application areas. It is envisaged that in the future these differences will decline when more and more technologies reach maturity. A critical factor affecting the direction of development is the standardization of technologies and services. Standards have worldwide high priority within the overall technological development strategy for information and telecommunication technologies. The main aims of this standardization policy are to secure the interest of the transport telematics-using sectors - now that effective deployment of transport telematics is one of the keys to maintaining competitiveness on national and international markets (viz. road transport operators), to remove barriers to trade in transport telematics, to ensure the viability of the transport telematics industry and to contribute to the creation of a competitive environment. By means of standardization policy, significant increases in the conformity with accepted standards for transport telematics products and services are expected. Worldwide, efforts are being made in this area, viz. the high number of institutions and organizations concerned with the establishment of standards (e.g. the International Organization for Standardization (ISO), the International Telecommunication Union (ITU) and the Comité Consultatif International des Radiocommunications (CCIR), and in addition national organizations).

In Table 2.11 an overview has been given of the expected development scenarios of transport telematics in the case of each distinguished functional area. Standardization has already been applied to a great extent in the area of *traffic information* for one-way transmitting traffic information systems. Internationally, standards for the use of the Radio Data System (RDS) have been set, and the launch of this system on the market has started. In the near future, the more advanced two-way transceiving traffic information systems will likely to be ready for the market. Individual traffic information systems will potentially be in large-scale production for individual motorists in the longer run.

In the area of *travel information*, some developments are already in an advanced phase, especially of those applications needing less advanced technologies. Various examples of relatively cheap but successful systems can be given, such as electronic railway guides, and videotex systems such as the French Minitel system. Systems providing integrated information on different transport modes, which require cooperation between the operators of different modes, are also in rapid development. In the near future it is envisaged that facilities providing such information and tourist information will be available in the street, at public transport stations, in offices and in the home.

In *public transport management*, the standardization process is slow. At present the technologies in development for operational management of bus fleets and buses themselves are diversified. It is here that international pressure for standardization is less since bus operators have a local or regional orientation. The adoption of technologies is likely to take place through local field trials of specific technologies.

In the area of *freight and fleet management*, the establishment of standards is of particular importance, since the potential user group is of a commercial nature. The setting of standards removes uncertainties with respect to the further development of systems which are a critical factor for the willingness of transport operators to participate in pilot tests. There is a big difference between non-mobile and mobile systems. Non-mobile systems are already in a late phase of standardization since their development started earlier. Standards for the electronic exchange of data have been well established. Recently, great efforts have been made to set standards for communication of data via satellites, due to the international and worldwide dimension of the use of satellites. A large-scale production is also envisaged here in the longer term.

In the area of *traffic and road management*, the pace of development has also been fast in recent years. However, due to the wide variety of telematics functions in this area and the site-specific implementation packages, it is difficult to forecast further developments. At present the production of applications is focused on stand-alone systems, while only in particular cases integration of various systems has taken place (e.g. the complex management of urban ringroads). In the future such integration will extend to the development of complete packages of systems for integrated road transport management.

Although considerable technological development has taken place in the area of *demand management* (e.g. smart card developments), until now the rate of development has been relatively slow in comparison with other transport telematics technologies due to the general low public acceptance especially of road pricing in many countries. In future and even in the short run, it is expected that at the political level more support will come, so that these systems will show an accelerated development line. It will however take several years to fully develop multilane/multi-use debiting stations, so that flexible road pricing will be possible.

As already mentioned in this chapter, technologies for *driver assistance and cooperative driving* in general are as yet in a relatively early stage of development. On the one hand, some specific tools assisting in the driving task have already been developed and are on the market (e.g. cruise control). On the other hand, systems are envisaged for automatic lateral control, which is only in an early phase of R&D in most countries. An exception is Japan where applications are being tested for lateral control and even the automatic following of routes by vehicles using extremely advanced technologies in both vehicles and road infrastructure. In Europe and the United States, such systems are not likely to be launched in the market within the coming decade.

The technological perspective of transport telematics set out in this chapter appears to be very large. Long-term implementation schemes of mature technologies for road and traffic management have already been defined by various governments (e.g. Rijkswaterstaat 1992a). However, as is the case with all technological innovations, the ultimate success rate among the public is dependent on the response of the potential user groups of the technologies. These user groups will be the subject of the next chapter.

Table 2.11. Technological development scenarios of transport telematics.

Application area	User type	Time horizon 1995	2000	2005
Traffic information	Motorists	• R&D • established standards & market launch of systems using one-way communication	• established standards & maturization of systems using two-way communication • market launch of systems using two-way communication	• large-scale production for consumer market
Travel information	Travelers	• R&D • diffusion among public primarily through public transport operators • market launch of private systems	• market acceleration	• large-scale production for consumer market
Public transport management	Public transport operators	• R&D • adoption through field trials by operators of not yet standardized systems	• established standardization • larger scale adoption	• (super)implementation of systems serving integrated functions
Freight and fleet management	Fleet operators	• R&D • established standardization of non-mobile telematics • market launch aimed at large operators	• established standardization mobilz telematics	• potentially large-scale production
Traffic & road management	Road authorities	• R&D • large scale production of stand-alone applications	• standardization of more integrated systems	• implementation of systems serving integrated road transport management
Demand management	Road authorities	• R&D • small scale adoption of precursory systems	• R&D • standardization of multilane/multi-use systems	• (super)implementation of multilane/multi-use systems
Driver assistance & cooperative driving	Motorists	• R&D	• R&D • market launch of stand-alone systems	• R&D • market launch of systems serving integrated functions

3 User Impact and Market Potential of Telematics

3.1 Introduction

In the previous chapter, a global view has been presented on the potential application areas and the range of transport telematics products already in existence or under development. We now develop the issues which determine the impact of transport telematics on the relevant actors in the market (the user impact) and the factors which determine the market potential. These two sets of factors are mutually dependent and will jointly influence such factors as take up, diffusion and the ultimate commercial success of transport telematics.

The aim of this chapter is to provide a framework within which to assess the user impact and the market potential for advanced transport telematics. We first present some insight into the roles of the various actors in the transport telematics market and the way in which they are likely to be influenced by transport telematics technologies. This is followed by an assessment of their response to transport telematics, based on two complementary conceptual models of travel behaviour, while next we will also explore the dynamics of the diffusion and awareness processes. Finally, a discussion is presented of the market potential of transport telematics, and of the range of possible barriers which may limit the effective impacts of transport telematics innovations.

The impact of transport telematics on users can manifest itself in three mutually linked questions:

- The *market* aspect; will potential users wish to have a personal transport telematics system at their disposal, owned or leased? Who are the potential users and how does information on transport telematics diffuse?
- The *level of use* of transport telematics; which factors determine the sophistication of the system and the level of detailed information required? How can information be both easy to access by the user and individually targeted to their own particular needs?
- The level of *active use*; which factors are important for adherence to advice and for the influences that transport telematics has on planned and actual decision choices?

A comprehensive analysis of the actors' roles and responses to transport telematics should deal with all these three questions, since they are highly interdependent. A decision to buy, lease or use transport telematics is dependent on a trade-off between the individual costs and benefits of the transport telematics system. If someone buys a personal information system, the expectation is that it will be used and that adherence to advice is of a sufficiently high level such that the value to the individual obtained from the purchase and the use of the transport telematics equipment exceeds the costs. This assumption seems plausible, as the purchase decision is normally based on individual rational economic behaviour, using detailed information from personal experience in day-to-day traffic bottlenecks faced. The purchase of a personal transport telematics system implies that someone's travel behaviour will not stem (entirely) from habitual factors, but that each travel decision will be based on the best available real time information and that this may change during a journey. Conversely, at the aggregate level, there is a relationship between the use of a personal transport telematics system and potential adoption rates within the market as a whole. The optimal level of market penetration will probably not be 100 percent, as the benefits to each individual user will then likely be reduced. The value of transport telematics systems stems from the behaviour of different users having access to different levels of information, so that the system as a whole operates at its most efficient level.

These questions will be dealt with in this chapter, which forms the methodological framework within which subsequent case studies can be positioned. In this sense, it acts as a structure for the case studies.

3.2 Actors in the Telematics Market

The telematics market comprises a large number of actors from both the public and the private sector. Four main groups of actors can be identified (Table 3.1), viz. the *supply* side, the *demand* side, an *intermediate* category and *external* influences.

3.2.1 Actors at the Supply Side

Actors at the supply side consist of producers of transport telematics hardware and software, including research and development. Examples include telecommunications companies diversifying from other sectors (e.g. from defence industries), micro-electronics companies, car manufacturers and software houses. Smaller companies, such as specialist consultancy companies, and university research institutions are involved in the design process. In most cases, applications are developed spontaneously. However, before a widespread application can be implemented, extensive discussions and agreements have to be organized with a vast range of external agents.

There are many good ideas in the area of transport telematics. But to gain public support and imagination and to guarantee market penetration is difficult, while the ultimate goal of commercial success is only achieved by a small number of ideas. Many innovations have consequences for transport policies and implementation requires action from the external actors (e.g. pro-active legislation prior to the introduction of road pricing would be required to make smartcards or metres installed in vehicles compulsory).

Similarly, many projects require external financing from public agencies in the research and development stage (including the EU) and from the private sector through venture capital funding. In each case the risks have to be assessed together with the internal coordination (agency) costs and external coordination (transaction) costs. The potential returns to investors are high, but so are the risks. It is important to be involved with innovation and new market products, but even with the large potential market for transport telematics products, there are relatively few market entrants.

Table 3.1. The main actors in the telematics market.

Supply Side Actors	Demand Side Actors	External Actors
Producers	Private Users	Government
Research & Development	Fleet Operators	Information
Consultancies	Collective Users	Finance
		Industry
Intermediate Actors - Infrastructure Owners and Managers		

The role of governments has changed very substantially over the past 15 years. The traditional interventionist approach with broad welfare objectives and notions of the public interest has been replaced by a neo-classical market driven approach. Previously, there was a balance between economic and other objectives, but now intervention is only seen as a means to compensate for market failure or for reasons of social justice. Most recently, the market approach has been moderated by broader concerns over equity and environment; so the competitive market is increasingly being replaced by the social market. In terms of transport telematics, this has meant that certain products have been seen as purely commercial with the competitive market determining the scale and nature of the innovation. Many of the logistics innovations and route guidance systems fall into this category (Chapter 2). Other products have been developed in the public sector and have been provided at public cost to improve the quality of public transport services (e.g. through public transport information systems) or the use of the road infrastructure (Table 3.3).

However, government still has an important and increasing role in the provision of services, in supporting research, in implementation, in setting standards and in regulating the transport telematics industry.

One great unresolved issue is that of information and control over the database. A tremendous amount of information is collected through the transport telematics systems from both the demand and the supply side. This information is crucial to the successful implementation of schemes as public acceptance and use of the information depends on its quality. Those systems which involve payment are even more dependent on that quality. Information is the new medium, but can it be left to the market? Its value is that it is real time information and can be customized to individual users. To achieve this immediacy and relevance is a costly process, but should this be controlled in the private or the public sector? There are questions which have not yet been addressed or resolved. It is unclear whether information and data is better collected and maintained in a regulated (public) or a deregulated (private) environment. If it is in a regulated environment, these are questions of responsibility and who pays for the information to be collected and maintained. Similarly, if it is in a deregulated environment, there are questions of regulation to ensure public interests are not compromised. Monopolistic control over information would place the owner of that information in a powerful position. Transport may be a special case, as the information has little commercial use, but the questions of ownership, responsibility and access need to be resolved.

3.2.2 Actors at the Demand Side

Actors on the demand side include private end-users, fleet operators and collective (or intermediate) users (Table 3.2.). In each of these cases it is argued that a significant improvement can be achieved through substantial user response to new forms of information supply. For example, in the passenger transport sector there are three main types of transport telematics systems where user response is important (Table 3.3).

Table 3.2. Demand side users of telematics applications.

Private end users	-	road users and public transport users.
Fleet operators	-	commercial and institutional operators of fleet vehicles including public transport operators, road freight operators, and forwarders, taxis, express service companies, emergency services.
Collective users	-	operators of publicly used telematics systems, including governmental bodies, road managers, parking managers.

From this classification it is clear that the new transport telematics technology lies at the heart of the innovation process of the transport sector, even though the direction and nature of causality and the degree of substitutability/complementarity (or synergy) are not exactly known. But it is evident that the transmission and movement of information are accompanied by drastic changes in the volume and nature of movements of people and goods, as is witnessed by the increasing popularity of on-line information and booking systems, video phones, video-conferencing, on-board navigation systems, and electronic data interchange. Thus there is a potential role for the transport telematics sector to upgrade the performance of transport systems in Europe. The field of transport telematics technology is obviously diverse and embodies many actors who in different roles will exert an influence on the adoption rate and the nature of new transport telematics equipment.

Users of the transport telematics technology will be inclined to attach a higher value to such equipment in case of congested networks, as can be seen from recent modelling experiments with the use of transport telematics information (e.g. Emmerink 1992 and Emmerink et al. 1994a). Individual users may make frequent switches as a result of transport telematics-induced road information; although this shift may increase the efficiency of a trip at the individual level, this does not necessarily lead to an increase in systems' efficiency (Emmerink et al. 1994). If full information is offered to all (potential) users, then the possibility of unstable systems' outcomes is rather high (depending of course on the sensitivity of individual trip makers to the information provided, in particular the quality of that information). This provokes the issue of overreaction and optimal access to transport telematics information. In addition, especially in case of on-board navigation systems an important socio-political issue may emerge. If only a small number of road users is equipped with transport telematics, then both these individuals and (probably) also the system as a whole may gain. On the other hand, if all users are provided with the same equipment and if all of them actually use it to such an extent that overreaction occurs, then the result may be a disturbance of the whole traffic system. Thus, full information use by all individual travellers may jeopardize the overall system's efficiency, unless the real time information provided is corrected for feedback effects of all behavioural responses. However, the current state of behavioural modelling is far from sufficient to account for such response changes.

Table 3.3. Classification of transport telematics systems where a direct impact on travel behaviour is expected.

1. **Traffic Management Systems**
   * Urban traffic control systems
   * Variable message signs - local warning messages
   * Fog detection and warning systems
   * Ramp control and speed limits
   * Monitoring of traffic and hazardous goods
   * Automatic Vehicle Location systems (AVL)
2. **Demand Management Systems**
   * Area access control
   * Road pricing and automatic debiting
   * Priority measures: automated car pool lanes, park and ride
   * Public transport information
3. **Driver Information Systems**
   * Radio traffic broadcast services (e.g. RDS - Radio Data Systems - to give local information)
   * Pre-trip planning systems (e.g. Auto Route - trip planning software to show best route)
   * In vehicle traffic information (e.g. Traffic Master)
   * Dynamic route guidance (e.g. Autoguide)
   * Communication of real time information to freight operators

In short, if all individuals have access to the same information at the same time, their own individual utility will be maximized, but the global effect might - and is likely to - be suboptimal. There may be strong arguments for different levels of information being given to different users (perhaps according to their willingness to pay). In this way, the individual utility can be improved subject to the overall constraint that the systems efficiency as a whole is improved.

User response is dependent on the type of trip being made and the type of technology available. The actual behavioural response may also vary - a road user may change route, departure time, destination, mode, and sequencing of activities. The availability of the different transport telematics systems gives the user a much wider range of potential options, and so the diversity is enhanced. By allowing more flexibility and complexity in behavioural responses, greater capacity can be squeezed out of a given infrastructure.

3.2.3 The Intermediate Category

Combining both the supply and demand side is the task of an intermediate category, the mediators. These are in general the infrastructure owners and managers (who may or may not actually own the infrastructure). In the case of private operation of a network, the manager will try to attract as much traffic as possible to maximize revenues. Congestion will affect revenues, so it is in the interest of the road manager (and hence of the road user) to cope with congestion by using transport telematics in some form. The same approach might be adopted by the public manager, but here the need for revenue maximization will be tempered by other considerations (e.g. environmental). The argument here is that better quality management of the road system and better information will mean that a greater capacity can be obtained from the fixed supply of infrastructure.

In most cases the actors are working together to provide the basic amenities to the public through the provision of the infrastructure, as well as by improving the performance of a given infrastructure through management systems (Table 3.3), including transport telematics.

3.3 Travel Choice Analysis and Transport Telematics

The role of the actors discussed above does not yet provide insight into the *individual decision chain* which determines the individual needs and use of more advanced types of travel and traffic information (see also Mahmassani and Chang 1987; Fischer 1993; Lotan and Koutsopoulos 1993). Modelling such individual travel behaviour is a difficult task and requires some critical assumptions on the basics of human behaviour (Stern et al. 1993). Such assumptions conform to the classic economic model of utility maximization where the decision to make a trip is followed by a trip planning stage which results in a ranking of the key trip characteristics according to individual utility before the choice is made (Banister et al. 1994). This choice may include all trip planning variables or a subset thereof (Figure 3.1). The assumptions made here include:

- rationality in choice: if the same choice set is presented again the same decision will be made (all other things equal);
- complete knowledge: the individual decision-maker has access to information concerning the accepted alternative together with knowledge of any rejected alternative.

Individuals act so as to maximize some benefit or utility function, with the individual being represented as exercising his or her choice over the full range of available options, limited only by constraints of time and money. In transport, this choice is normally represented as a discrete alternatives between mutually exclusive modes, but it can also be used to analyze other parts of the travel decision. The addition of transport telematics information systems (e.g. on public transport, VMS or route guidance) reinforces the critical role of the information assumed in the utility maximization model, as decisions will be based on the best available information prior to the trip being made and during the actual trip.

An alternative model would argue that decision makers are not utility maximizer but satisficers (Banister et al. 1994). The individual makes choices in a situation of partial knowledge. When certain thresholds are reached (e.g. significant foreseeable delays), action will take place. Such a procedure explicitly involves feedback with the result that each new trip may be modified by previous experience, which could be positive or negative (Figure 3.2).

Such a model gives rise to a more complex decision process and would suggest a more selective use of any pre-trip or in-trip transport telematics information, as all information is modified by previous experience. In turn, this would suggest that transport telematics information has to be selective and targeted to individual users, as more general information may not be relevant.

Information given to transport telematics users would also have to be accurate, as any failure in the system would result in the strengthening of the individual's own experience as opposed to the experience from the information system. Any such reduction in the quality of transport telematics information would reduce reliance on it, the market for it, and the price that would be paid for it. Selective use of transport telematics information is a key research area about which little is known.

Information can either be provided at no direct cost to the user (Type I) in which case utility will be increased, or at a cost to the user (Type II) in which case the increased quality and value of that information will have to be balanced against the cost of that information. Similarly, on the supply side, information can either be provided to users in general (e.g. VMS and public transport information systems) or to the user on an individual basis (e.g. route guidance). Examples of pre-trip and in-trip information product types are shown in Table 3.4.

Given these different user and supplier constraints, there will also be differences in levels of adherence to advice. Using the utility maximizing framework with no direct cost to the user, utility should be increased with a greater use of both pre-trip and in-trip information. If there is a direct cost, then there will be a situation where no use of transport telematics is made (no change in utility) or where there is use of pre-trip and in-trip information, thus increasing utility, either will full adherence to advice or partial adherence to advice. It is here that the changes in utility may be difficult to assess as well as the conditions under which advice is accepted or rejected.

Using the satisficing behaviour framework, some further differences may be observed. If the trip is a new one, then behaviour is similar to utility maximizing, but if the trip is one which has been made before, then satisfaction and feedback become important. Selective use will be made of transport telematics if there was dissatisfaction with the previous trip, but if that trip was perceived as satisfactory, then there is no need to have any new information. Again, adherence to advice can be either complete or partial, depending on the type of information required, previous experience and levels of satisfaction. This is where concepts of experience and familiarity become important - behaviour may become routinized.

In both models, the determinant factor in the success of different forms of transport telematics depends on the quality of the information and when and how it is presented. In-trip information has to be more general in the level of advice given and is limited by the size of the display available. To expect major changes in behaviour resulting from such information may be optimistic, as the information is not only relevant to that particular route. Alternative routes are not given on VMS or public transport information systems. Unless the traveller has a good knowledge of alternative routes (the assumption in utility maximization), the perception might be that there is no choice and so the action would be to remain in the queue at the bus stop or on the road. The benefits of such systems are that they are provided free at the point of use and they are likely to raise levels of satisfaction with the service being provided: a reassurance utility.

There is a much greater potential with respect to pre-trip information where options and choices can be made on a much wider variety of behavioural variables. It is here that decisions on trip timing (including trip scheduling or trip suppression), alternative modes, different destinations and the best route can all be made. The range of information and the personalization of the relevant parts of that information can be made available to the user, so that the assumptions of rationality and knowledge can be met. However, this requires people to spend time prior to making a trip extracting the relevant information. This again involves a cost in the time spent, and may only be appropriate when a new or exceptional trip is being made or when problems are expected. Similarly, many people may argue that their degrees of freedom in terms of mode, destination, start time and even route are limited. Route guidance systems attempt to address some of these problems, but again the information given relates to route choice, not to the other factors which make up the trip. For example, it does not offer the opinion to suggest to driver to park the car and take the train to the destination, or to give a range of alternative destinations (e.g. shops).

It seems that there are a series of major conceptual issues which need to be thought through on transport telematics systems. To some extent they depend on the theoretical model being used, but more generally they relate to the type of information being given to the traveller, when and where the information is given, and the relatively narrow range of options available to change behaviour. To expect major behavioural changes resulting form the types of transport telematics currently being developed may be optimistic. With these limitations (e.g. on measurement and change) in mind, it is still meaningful and possible to set up experiments and collect data for analysis and evaluation.

The difficulties arise when the actual nature of the behavioural change has to be specified and the usefulness of the information to the users has to be quantified. It is here that the range of options seems to be limited, both in terms of the information provided and the possible user response. Type I applications (Table 3.4) give the same information to a large number or users, but give no indication of the choices or alternative actions which could be taken. Type II information relates to the specific user, but only gives a limited range of alternative (e.g routes). Even here, the usefulness of the information may be limited, particularly if the user is familiar with the local road network.

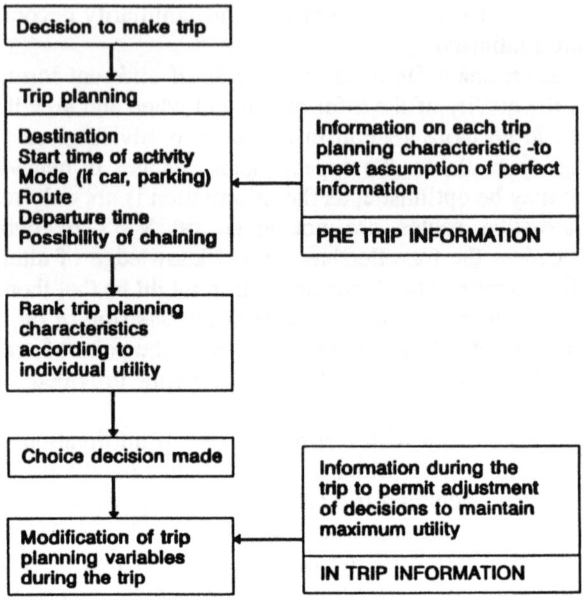

Fig. 3.1. Decision processes based on utility maximization.

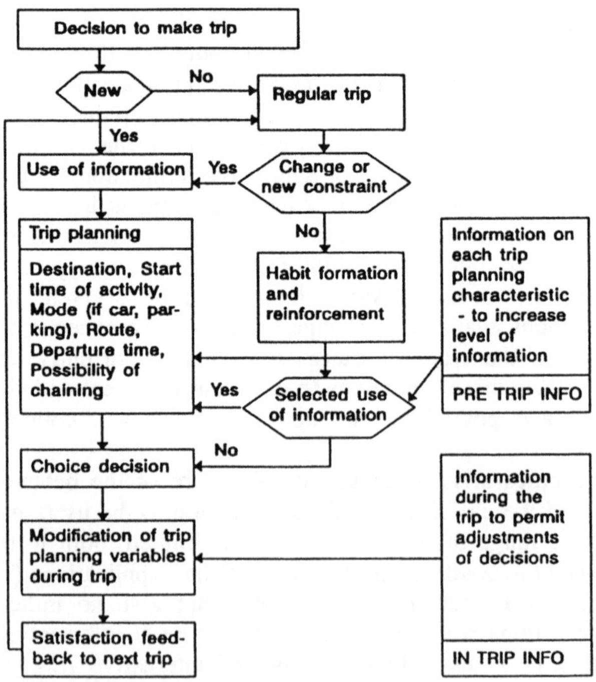

Fig. 3.2. Decision processes based on satisficing behaviour.

Table 3.4. Examples of pre-trip and in-trip information supply.

PRE-TRIP

	Type I - No cost	Type II - Cost
General	CEEFAX ORACLE	In home terminals PROMISE MINITEL
Particular	Timetables	Route planning Telephone enquiry

IN-TRIP

	Type I - No cost	Type II - Cost
General	Variable message sign Passenger transport information system	AUTOGUIDE TRAFFIC MASTER
Particular	Radio data system	Route guidance

Two different decision making structures have been presented, one based on utility maximization (Figure 3.1) and the other on satisficing behaviour (Figure 3.2). It seems that there are natural limits to the range and detail of information which can be provided during a journey. Much more detailed information on a range of choices can be obtained prior to the trip being made. Only in the home or in the workplace or shop can the full range of options be assessed.

We must recognize the limitations on the range and magnitude of responses to transport telematics, at least at a theoretical level. In a practical sense, there is still a need for an analysis framework which can link the transport telematics to user response. The types of changes that will take place will not be universal or large scale, but varied and specific to particular user groups. Changes will relate to behaviour, to attitudes and perceptions, as well as levels of satisfaction with services provided.

3.4 Actors' Response to Transport Telematics

3.4.1 Introduction

The assessment of transport telematics user impact is both fascinating and complex, and implicitly linked with awareness, the dynamics of diffusion and market potential. The following schematic diagram gives a flavour of some of the interdependencies (Figure 3.3). This process varies between the different types of transport telematics technology, while the individual response depends on the characteristics of the user and those provided at a cost to the user. As we shall see in the case studies (Chapters 5-8), young people, particularly young males, are likely to respond positively to innovations in general, and transport telematics in particular. Secondly, an important distinction must be drawn between transport telematics systems provided publicly at no direct cost to the user, and that provided either publicly or privately with a charge to the user. In the first situation, there are two main demand actors, namely the operators of the systems and the end-users. Examples here are public transport information systems, and parking and other road information provided on VMS Systems (Table 3.4). Investment decisions of operators of these systems are based on the likelihood of being able to influence the attitudes and behaviour of travellers, and the consequent beneficial impacts on the efficiency of the transport system as a whole, in terms of less congestion and better quality public transport. For the second situation, the only demand actor is the private end-user. The transport telematics applications include in-car systems and home-terminals providing travel information. The market viability of the transport telematics depends on the potential end-user's attitudes and perceptions towards these systems and their own requirements. Use of these systems may relate both in their utility in terms of rational use and their status in terms of their novelty value.

Equally important as the introduction of a new technology is the dynamic process of innovation diffusion (cf. Bertuglia et al. 1995) (see also the schematic representation in Figure 3.4). Crucial to this vision is the argument that the commercial diffusion of transport telematics technologies is dependent upon the growing awareness levels of the relevant actors involved with each application. The issue of awareness raising and targeting particular sections of the population are the two key ingredients to market potential (Section 3.5). Impacts on individual travel behaviour can be *direct* where the intention of the transport telematics is to change actual decisions (e.g. route guidance). Alternatively, the impacts can be *indirect* where the intention of the transport telematics is to improve the quality of decisions and more generally the quality of travel (e.g. driver assistance facilities and public transport information).

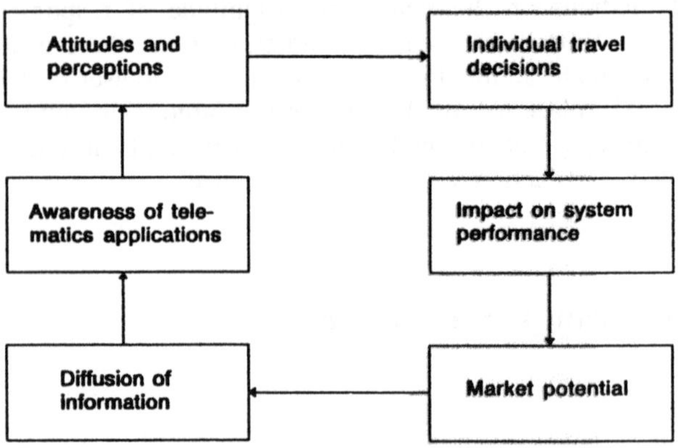

Fig. 3.3. The interdependencies between transport telematics and the complexities of user response.

The functional classification developed in Figure 3.4 divides the diffusion process into three component parts - the impact, the means of implementation and the actors involved. The following three tables elaborates on each of these in greater detail (Tables 3.5 - 3.7). In each case the six groups of transport telematics applications developed in

Chapter 2 are used to indicate the links between the component parts. In Table 3.5, it is clear that it is in the interest of the system's manager to raise the awareness of the public, particularly where the effect is in the public domain and area wide (e.g. access control in cities). Similarly, the transport telematics applications can be market driven or public policy driven (Table 3.6). The range of actors on the demand side and the intermediate side are summarized in Table 3.7 and are used here as in Section 3.2 where the main groups of actors were presented (Table 3.1).

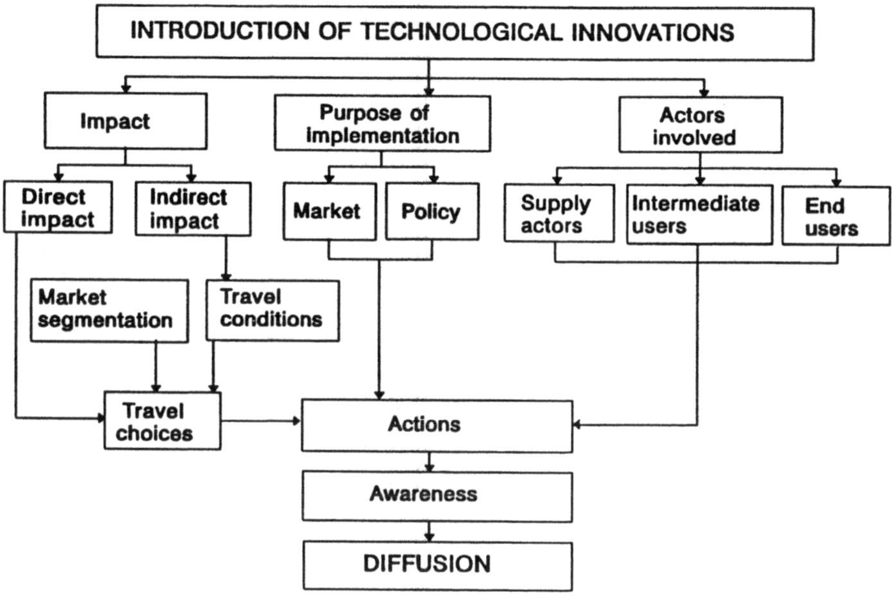

Fig. 3.4. Determinants of the diffusion process of transport telematics.

Table 3.5. Technological innovations and the impact on travel behaviour.

Group of telematics	Impact on travel behaviour
Demand management Public transport management Traffic management Travel & traffic information	Direct
Driver assistance Freight and fleet management	Indirect

The interactions of these three components of this diffusion process feed into travel choices and other actions, and the gradual rise in awareness leads to a further round of diffusion. This is the dynamics of the process (Figures 3.3 and 3.4). Throughout this process, certain elements reappear and these have been combined into a conceptual approach for the interpretation of the social impact of transport telematics - the *Nested Approach*.

Table 3.6. Technological innovations and the means of implementation.

Group of telematics	Means of implementation
Demand management Public transport management	Policy driven
Traffic management	Policy and market driven
Freight and fleet management Travel & traffic information Driver assistance	Market driven

Table 3.7. Technological innovations and the actors involved.

Group of telematics	Actors involved
Demand management Public transport management	Intermediate users End users
Traffic management	Supply actors, Intermediate users, End users
Travel & traffic information Driver assistance Freight and fleet management	Supply actors End users

3.4.2 The Nested Approach

In the complex and dynamic process which will ultimately lead to the diffusion of transport telematics as outlined before, a critical success factor will be the social acceptance by the end-users. In the field of assessing the social impact - including changes in travel behaviour - which will result from the introduction of transport telematics, relatively little conceptual analysis has been undertaken. Such investigation requires a coverage of a wide range of relevant (dynamic) behavioural issues. A comprehensive (nested) approach to deal with all relevant aspects should include the impacts of a full range of areas of operational interest in telematics (BATT 1992). The benefits of applying such a comprehensive approach would be:

- to provide behavioural parameters to those involved in the development of telematics systems;
- to inform authorities of the best ways to implement the new technologies and to increase the success of transport telematics technology;
- to inform the industry of better ways to promote and market transport telematics technologies to meet user requirements.

The nested approach integrates the elements described above allowing the impacts of telematics to be measured at three separate *levels of reference*: first, at a strategic level, secondly, at the level of market potential, and thirdly, at the level of market responses.

At the *strategic level*, the concern is with the overall system wide impacts, given certain types and certain levels of introduction of transport telematics. Within this framework, various changes can be assessed in terms of user and producer benefits, the direct and indirect environmental impacts, reductions in accidents, energy savings and the use of the infrastructure. The assessment may cover total performance of the system, the distribution and equity implications and the technological achievements.

At the *market potential level*, the concern is with the means by which the potential for transport telematics can be maximized in terms of acceptability and penetration to the various parts of the market. Part of this marketing is to access market awareness of the product, while the other part is to identify which segments of the market are likely to represent the greatest potential for telematics. It is realized that some people will be more positive about the use of telematics than others, and that not all people will use it in the same way. The identification of different markets is a very important part of applied research and will provide a link between research and the telematics industries.

At the *market response level*, the concern is with the costs of the technology, changes in individual behaviour and the scale of implementation. The focus here is on the cost effectiveness and the direct benefits to the individual users of transport telematics, the range, scale and timing of introduction of transport telematics and the rate of behavioural change which might follow. Much - often hypothetical - research has already been carried out on the impact of telematics, often in terms of the most optimistic scenario if saturation of the technology is achieved over a very short period of time.

Table 3.8 brings together these three levels into a composite table. The three levels of reference are put off against three categories of evaluation criteria, namely technical, socio-economic and political/dynamic criteria. The cells of this table contain the main areas of investigation.

Table 3.8. Head elements of a nested approach.

	Levels of reference		
Evaluation criteria	**Strategic**	**Market Potential**	**Market Response**
Technical	Performance	Marketing	Cost of technology
Socio-economic	Distribution & Equity	Segmentation	Behaviour
Political & Dynamic	Technological Perspective	Awareness	Diffusion

It should be noted that not all telematics applications are likely to feature in each element of this framework. For instance, traffic information and public transport information will be evident in all cells, but other telematics applications, (e.g. demand management) will only relate to market response, since it is system wide and affects all users.

Given the emphasis on the end-user side, some elements of this table need a further refinement: *segmentation*, *behaviour*, *awareness* and *diffusion*. Here the focus is on the behavioural response in combination with segmentation factors that are necessary to establish the market potential. These four issues are now elaborated.

Behaviour

The argument here is that user behaviour will change as a result of the introduction of transport telematics, but that changes may vary according to the individuals, the situation and the type of transport telematics being tested. A range of behavioural responses has been identified which might be anticipated for a particular journey at one point in time. These include: mode shift, departure time, change in route and destination, trip generation/ suppression, trip scheduling, parking choice and adherence to advice (Table 3.9). They are elaborated below:

Mode shift: The impact of transport telematics may result in users shifting mode in order to gain time or to save costs or meet their constraints.
Departure time: A shift may occur in departure time, given the transport telematics information on the current level of congestion or the generalized cost of the prospective trip if individual utility is to be maximized or to meet specified preference constraints.
Route: Provided that the technology is available, route choice may be modified. Route diversion or adherence to advice supplied by transport telematics may be influenced, not only by items cited in the segmentation (such as familiarity), but also by reliability of the information provided.
Destination: Decisions may be made to select alternative destinations if the route previously selected is congested or if the transport telematics system can give information on alternative opportunities. Destination choice is clearly relevant to some types of discretionary trips.
Trip generation/suppression: Technology may influence the decision whether to make a trip or not, as advice on congestion may result in trip deferral or cancellation within the decision period considered. Purpose and need of the journey will be decisive in establishing possible changes with respect to whether a trip is cancelled or not.
Trip scheduling: This arrangement within a determined user dependent time period may be considered if satisfactory or non-acceptable alternatives are suggested by transport telematics (e.g. route, trip timing, parking).
Parking choice: Parking decisions may be influenced by access to transport telematics information regarding the location of car parks, the availability of space, and the advice on the route to be followed.
Adherence to advice: Adherence to information provided may be influenced by many factors such as the quality of the information being provided, reliance on such information, familiarity with the network, previous experience and user characteristics.

Segmentation

Here the concern is with the main socio-economic characteristics of the individual which might influence both the decision to acquire a particular form of transport telematics and the actual use of that transport telematics at any point in time. The argument is that not all people require access to the same technology and that even if they would have that technology, use patterns will vary. Meaningful segmentational factors would include:
Car availability: The availability of a car identifies not only social groups, but also provides information such as existence of alternatives or dependency on the public

mode. It may be important to select segments of both groups. This segmentation has proved to be useful by many previous studies.

Age/gender: It can be expected that the penetration of transport telematics may be differentiated by age and gender of potential users. Younger people may be more likely to respond to innovation than older people and men may be more responsive than women.

Social group: Socio-economic group, type of employment and some measure of class may all affect patterns of use of transport telematics, both in terms of actual takeup and in terms of marketing.

Income levels: Income is closely related to the social group and it is likely to be the main factor in the decision to acquire the transport telematics technology or to obtain access to it.

Experience: Positive or negative experience may modify the usage of transport telematics. Past experience is traded off against transport telematics-supplied information on the current situation. Increased reliability of current information must be maintained and improved. This factor is related closely with user familiarity and awareness of the alternatives available.

Familiarity: The issue of familiarity has been identified in previous research as important in determining whether pre trip information is required in home (unfamiliar trips) or during the trip (familiar trips).

Purpose: Trip purpose may also help to identify which types of activities have the greatest potential for transport telematics. Discretionary trips (e.g., social, leisure and shopping) may present greater opportunities than regular trips (e.g., work and education) where there is a much greater degree of familiarity.

Table 3.9. Important market potential and market response parameters.

	Market Potential	Market Response
Socio-economic	SEGMENTATION 1. car availability 2. age 3. social group 4. income group 5. experience 6. familiarity 7. purpose	BEHAVIOUR 1. mode shift 2. departure time 3. route 4. destination 5. trip generation /suppression 6. trip scheduling 7. parking choice 8. adherence to advice
Political & Dynamic	AWARENESS 1. exposure to ATT 2. acceptability 3. publicity	DIFFUSION 1. pre-conditions 2. take-off 3. saturation levels 4. adaptation

Awareness

Innovation takes time for people to become aware of, while awareness often relates to exposure or experience. Part of that process is publicity, but equally important is the public acceptability of innovation, the perceived necessity and the benefits. Some relevant issues are:

Exposure: Previous knowledge and exposure to the technology may be decisive in the usage of any future application. This exposure relates to knowledge, experience and acceptability, as well as to user characteristics.
Acceptability: Apart form awareness of technology, there is a considerable problem concerning the public acceptability of technology (e.g. the debate on road pricing and privacy). Innovation takes time to become accepted and the market response may be seriously affected if social factors are not positive towards it.
Publicity: Awareness and acceptability can be raised by publicity and marketing which will both promote transport telematics technology and help to allay any concerns that people may have.

Diffusion

Innovation diffusion also takes time as the market does not respond instantaneously. Even when all conditions are favourable, responses have to be monitored and evaluated over a significant period of time, as standardization becomes possible and substantial economies of scale prevail. Critical diffusion parameters are:
Pre-conditions: These are the necessary political and technical conditions which have to be in place prior to any large scale application of transport telematics and relate to a willingness to address environmental and traffic problems.
Take off: As diffusion takes place, initial interest begins to snowball and market penetration expands at a faster rate after reaching a critical acceptance threshold which depends largely on the conditions.
Saturation levels: With maturity, a saturation level is reached, but suppliers then identify new markets to ensure the total market for all transport telematics continues to expand.
Adaptation: The closely linked dynamic process outlined in the move from pre-conditions through take off to saturation is not a unidirectional process. There are also important feedback effects, as individuals and companies modify their behaviour patterns and change habits.

3.4.3 Conclusions

User response to transport telematics is a complex process, dependent upon the exact nature of the response (behaviour), who is responding (segmentation), the means by which the system has been advertised (awareness) and the dynamic process of diffusion. In many cases the range of the response is large and the scale of response may be small. Not all changes in behaviour and market responses will operate in the same direction, hence the requirement for careful and systematic survey and analysis (Chapters 4-8). The importance of understanding the nature and scale of user response is critical to the success (or failure) of transport telematics systems. The technological arguments need to be matched by the appropriateness of the interface with the user. Equally important is the actual use made of the technology. People and firms are naturally resistant to change, but are interested and curious about a new technology. They are willing to try, but if they receive a bad experience for whatever reason, their reluctance to adopt new ideas is increased. Often only one or two trials are given. Secondly, as noted in the Nested Approach, the scale of the change is not likely to be great and it is likely to vary between respondents. The expectation that a substantial change in user behaviour will take place (e.g. through route choice or through modal split) is unrealistic. The measurement of change is a complex empirical exercise. Related to the question of the nature and scale of response is the market potential for transport telematics systems. These two factors combined form the main components of the Nested Approach described in this section (Table 3.8).

3.5 The Market Potential of Transport Telematics

3.5.1 The Cascade System

An assessment of the market potential of transport telematics from the view point of its use and relevance (including its adoption and commercial introduction) requires a consistent treatment of all necessary steps ranging from the identification of inefficiencies in transport networks to a final assessment of its market potential from an industrial perspective. The way in which this series of steps can be described in a structured and consistent way is based on a so-called Cascade System. Following sequentially all steps of this Cascade System will ultimately lead to a final assessment of the potential of the transport telematics market (MARTA 1993a). In general, a cascade forms a series of steps in which the output of the one step is at the same time the input for the next step. Such a chain of connections requires a careful analysis of linkages in this multi-step approach using in a pairwise manner a series of relevant dose-effect relationships. In this way a systematic impact assessment methodology can be developed.

In the architecture of the Cascade System seven building blocks modules are distinguished (see Figure 3.5) which altogether form a logical and applicable series of steps for assessing the transport telematics market potential in urban areas and along inter-urban corridors. It should be noted, that in various stages of the Cascade System, feedback loops may exist. A comprehensive representation of all sequential steps in the Cascade System, starting from the first building block and continuing to the last building blocks are presented in this section. By using such systematic design principles, a comprehensive and practical research framework has been developed.

Module 1 - Transport Inefficiencies and Transport Networks

The first stage in the chain is an assessment of relevant European road transport networks or corridors which are particularly suitable for transport telematics applications. The assessment consists of an effect table comprising the relevant evaluation indicators or criteria for identifying, evaluating or selecting transport telematics-relevant transport networks or corridors. Indicators of transport inefficiencies would include the existence of bottlenecks, landscape value, accident and fatality rates, average speed and the ratio between passenger and freight traffic. In the urban situation on street parking availability, accessibility to public transport and the ratio between private and public transport could be added to the list. Multi-criteria techniques would be used to establish where the main inefficiencies have occurred or are likely to occur. It would be these corridors or networks where there might be the greatest potential for the application of the transport telematics technology.

Module 2 - The Transport Networks and Technology Module

For each of the transport networks identified in Stage 1, the range of transport telematics applications is considered in terms of their relevance and suitability for use in tackling the network inefficiencies. This module contains the linkage structure between various transport telematics-relevant features of corridors (e.g. bottlenecks, congestion and pollution) and the various transport telematics technologies which serve to alleviate proven situations on these corridors or networks. Examples of transport telematics technologies would include electronic documentation, automatic debiting systems, tracking and tracing systems (e.g. Automatic Vehicle Location), route guidance systems or trip planning information. The full range of technologies has already been presented in Chapter 2. Empirical research, together with desk studies, is required to identify the specific alternatives provided by each transport telematics technology to address the problem in the transport corridor or network. Technology assessment methods can be used to make a meaningful judgement of the transport telematics technologies.

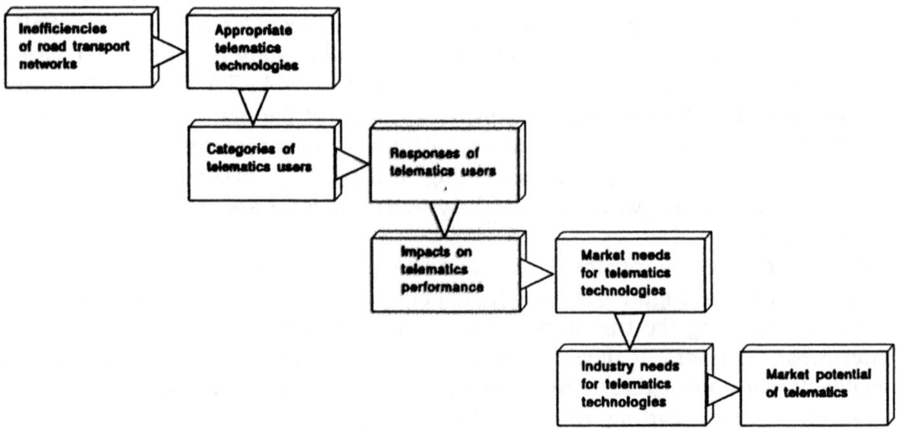

Fig. 3.5. The Cascade System for assessing the market potential of transport telematics.

Module 3 - The Technology and User Module

This module forms a bridge between various classes of potentially available and suitable transport telematics technologies on specific types of transport corridors and networks, and the functions (benefits, costs and other advantages) of these technologies for various user categories. This module includes the dose-effect relationships between the characteristics of various types of relevant and feasible transport telematics technologies of given types of transport networks. It also assesses the importance of each technology for the individual user categories, including road users (e.g. commuters or tourists) and corporate actors (e.g. freight hauliers or public transport operators). Each of these user categories has its own interests and priorities, and hence attaches a specific value to a given type of transport telematics technology. It links in closely with the market response and market potential categories identified in the Nested Approach

(Table 3.8). Information on this module is generated from the field trials, surveys, desk research and discussion with transport telematics experts, including the industry.

Module 4 - The Use and Response Module

This module incorporates information on the responses of transport telematics users for each distinct relevant transport telematics technology on each specific transport corridor or network. This module is at the centre of the Cascade System and again links closely with the Nested Approach. It aims to make a systematic inventory of types of responses of various transport telematics users. Examples of user responses include route change, change in departure time and cancellation of trip (Table 3.8). These responses may differ between transport telematics users and one individual's responses may change over time. For example, freight hauliers might decide to adopt a new system of logistics platforms as a result of the introduction of transport telematics technology, while commuters might decide to change route or mode. There is very little data available on user response, and extensive field work is required for the assembly of consistent and comparable information.

Module 5 - The Response and Performance Module

The various response categories (e.g. changes in route choice, departure time and rescheduling of activities) are linked to measures of performance of the system as a whole. This would include savings in travel time, improvements in safety, reductions in pollution levels and more general social benefits. Although the conceptual structure of this module is clear, there is a difficulty in attributing cause and effect. As noted before, the impact of transport telematics is diverse and often small scale and any aggregation into an overall assessment of performance is difficult as other changes may be taking place in the system. For example, increases in fuel prices or public transport fares may also influence user response and the performance of the system. A systematic impact assessment needs to be designed, based on a with-without comparison with external factors remaining constant. Surveys are required with the key actors and controlled experiments to isolate the impact of the transport telematics technology.

Module 6 - The Performance and Needs Module

This module translates the improvements in system performance into an estimate of the market for transport telematics technology, given the variations in locations (corridors and networks) and the size, range and quality of the applications. This stage is an essential part of the Cascade System, as it provides an assessment of the expected demand (or needs) for various transport telematics technologies. As a result of performance improvements of transport corridors and networks, various marketing research strategies, based on consumer and user behavioural responses, have to be developed. The module forms an interface between industrial (or producer) interests and user interests. Apart from the information from field trials, key actors need to be interviewed. It is likely that there is (or should be) a feedback process to the previous two modules or that there is a dynamic process of diffusion taking place. Interviews, surveys and expert knowledge collected at several different points in time are required, and it may be appropriate to have a panel or longitudinal survey to isolate the changes in user response, performance and needs over time.

Module 7 - The Needs and Industry Module

The final module in the Cascade System translates market needs to industrial market potential for the full range of transport telematics activities. The expected industrial market implications resulting from the needs for different types of transport telematics technologies on corridors and in networks are recorded. Subdivisions into the relevant branches of the information and communication technology sector have to be made. Again, the information has to be obtained from interviews with industrial specialists, the transport telematics operators, transport telematics producers (e.g. the micro-electronics industry) and other transport telematics experts. It is largely the industrial research which is at present unknown.

3.5.2 Concluding Remarks

Appropriate application sites and appropriate transport telematics technologies, the assessment of which constitutes the first two modules of the Cascade System, are dependent on objective factors (i.e. factors related to available infrastructure, transport logistics and technology). The assessment of the other five modules are based on subjective factors (i.e. factors related to the demand side of the transport telematics market), which constitute the social impact of telematics systems. The links with the short term impact on users as outlined in Section 3.4 is clear. This section has developed the more particular aspects of user response and has placed it within the broader framework of transport policy analysis. The problem of network inefficiencies, together with the identification of key problems which transport telematics can help resolve, are two essential pre-requisites for applications. Similarly, the market potential in terms of user response and the transport telematics industry needs to be covered, if the commercial potential of transport telematics technologies is to be fully realized. In this section we have expanded the narrow view of the actors response to transport telematics taken in the Nested Approach and placed it in the wider dynamic framework of the Cascade System.

3.6 Conclusions

3.6.1. Introduction

This chapter forms the major conceptual framework against which the impact of Advanced Transport Telematics (transport telematics) can be assessed. It has described the roles of the various actors in the transport telematics market and the ways in which they are likely to be influenced by the technologies. Travel choice theory and behavioural response have then been presented, both in its traditional sense and in the modified version where new forms of information and choices are presented to the travellers. The discussion has also covered the important issues of market segmentation and diffusion of innovation. All of these factors will have a direct impact on market penetration and the commercial viability of the transport telematics technologies.

In this conclusion we attempt to provide an overview of some of the key unanswered questions which are taken up in the subsequent chapters on the transport telematics case studies. Two fundamental points need to be raised here. First, the technology is varied and wide ranging in its applications. This comment relates both to the type of technology and to the role of the different agencies in research and development. Some transport telematics technologies will be primarily part of the public sector provision of services, particularly information services. Other transport telematics technologies will be provided by the private sector and these mainly relate to in-vehicle systems where the user will pay directly for them. A third group of transport telematics technologies will affect companies in the organization of their services or distribution networks. A fourth group is more appropriate for inter-urban applications rather than urban applications, with spatial scale becoming more important. Table 3.10 gives some indication of the range of applications, together with the chapters in which they are taken up. This figure does not claim to be comprehensive, but it does give a flavour of the rich potential for transport telematics applications. Similarly, these technologies are not exclusive and the same technology has many different applications. The case study chapters will give further details on the technology and user responses from applications in several different countries.

Table 3.10. The family of transport telematics applications.

PROMOTION	TYPE OF TRANSPORT TELEMATICS			
Public sector	Passenger transport information systems (Chapter 7)	Demand management	Variable Message Signs (Chapter 6)	Integrated urban traffic management
Mixed	Parking guidance systems	Management and information systems	Road network information (Chapter 6)	
Private sector	Finance and organization of companies (Chapter 5)	Route guidance systems (Chapter 6)	Freight distribution and management systems (Chapter 8)	Driver assistance and automatic control in driving

The second issue concerns the ownership and control of the databases. Transport telematics applications allows a tremendous amount of information to be collected about the operation of all transport systems. The value of that information relates to its quality and the uses to which it is put. Information is collected on a real time basis from users of the system, from the system itself and from outside agencies. It is only when all elements are put together in a real application that the full potential for transport telematics is realized. Several unresolved questions are raised by the advent of the information era. Should the ownership of the information be in the private or the public sector? Who has responsibility for the database and who are they accountable to? How has access to the database to be controlled and what safeguards are there for all users? The information revolution has raised these issues, not just as they relate to transport, but in many other sectors as well. The successful implementation of many of these systems depends on the quality of the information, and this in turn relates to the control of the databases. Power may shift from the economic base to the information base (Banister 1994a). Figure 3.6 attempts to present some of these questions. These organizational issues, together with accountability and responsibilities, have yet to be resolved. This is a necessary requirement prior to widespread application of transport telematics, but it is also important for the small scale demonstration prospects which have been mounted.

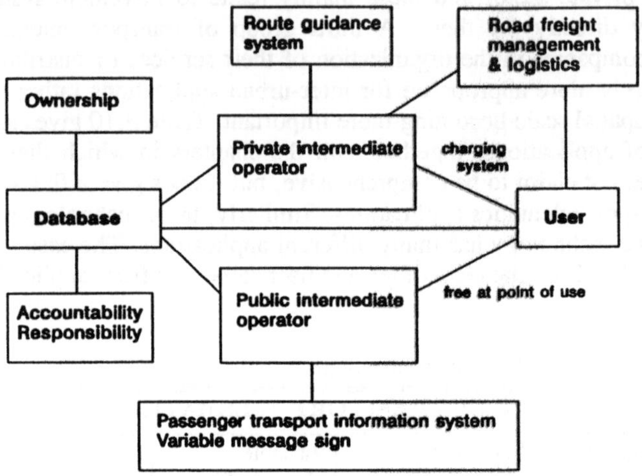

Fig. 3.6. Linkages between the database and users.

In addition to these major questions, there are a series of other questions which also need to be addressed. In this conclusion we list them. They will subsequently be taken up in the case study chapters and we will return to them in the conclusion to the book.

3.6.2 Some Open Research Questions

The following research issues need - in light of the above observations - due attention in the future.

Full evaluation of the costs and benefits

The transport telematics market is vast, both in terms of its potential applications and in terms of variety of actors involved. Where does the greatest potential actually lie? (Table 3.11). Little research seems to have been carried out on estimating the market potential, either in a commercial sense or in terms of the greatest need. Many of the projects implemented have not carried out a full evaluation of the costs and benefits (cf. Nijkamp and Blaas 1994). In many cases a conventional evaluation may be inappropriate, as the scale and range of impacts is likely to be great and vary substantially between applications. Yet the successful implementation of transport telematics is crucially dependent upon sound economic analysis.

Market for transport telematics

Related to the above issues is the absence of any real understanding of the market for transport telematics. Not all users will respond in the same way to the information and the market may be quite limited, at least initially. The market will depend upon the perceived need and usefulness of the information being provided, the presentation of that information and most crucially the use to which it is put. The question of interpretation of information and adherence to advice has not been addressed. It is often assumed that individuals are utility maximizers and that they are trying to minimize their generalized costs of travel, usually by taking the shortest path. But, as we have seen in this chapter, other theoretical models may be equally appropriate. Individuals may be satisficers looking for key thresholds in terms of acceptable times for travel. Alternatively, on some journeys, time may not be important.

Willingness to pay

Willingness to pay for information has not been tackled, as demonstration projects have been provided free at the point of use. Yet there are costs (some substantial) of setting up transport telematics projects and in collecting the information database. The evidence available from stated preference research suggests that there is a very low level of willingness to pay, both for information and for access to otherwise congested locations (e.g. road pricing). This reaction is not unexpected, as users already feel that they already pay too much for travel, and they are not clear as to what additional advantage they will receive from paying more. But, if transport telematics is to prove successful, then users will have to pay for the information provided, particularly if it is provided by the private sector.

Quality of the information

Related to the willingness to pay issue is the quality of the information provided. New ideas take time for diffusion and many people and organizations are prepared to try the new technology, but their attitudes and reactions will be strongly determined by the quality of the information provided. In the past, information has been provided to all users in a 'blanket' format. For example, one may just look at a timetable and extract the time of the train which one wants to take. Transport telematics provides the opportunity to customize information to the individual user, so that the search process for the precise piece of information required is done for him or her. So complex travel arrangements or routes can be determined with the help of the information system. Conversely, poor or unreliable information provision will lead to a rejection of the technology. Quality depends on both the information and the service being provided. It may be appropriate to use sometimes qualified experts as an intermediary to provide the information like in the case with telephone hotlines. Expert systems might also be used in this context.

Passive systems

The active involvement of the user in the acquisition of the information seems the most attractive option. The user requests the information and then reacts (positively) to it. However, there is also tremendous potential in passive systems about which the end user has little knowledge. Technology has a major role to play in the management of vehicle engines, in the management of traffic systems and in locating incidents in the network. Initially, the greatest potential for transport telematics applications may be in these management activities rather than in applications which require a direct response from the user.

Table 3.11. The road transport telematics market.

Market segment	Market nature	
	Private/Commercial	Public
Individual road users	Personal car drivers (business drivers, commuters, tourists etc.)	
Operators of vehicle fleets	Road freight operators Bus operators Taxi operators Operators of express services	Mobile police Fire brigades Ambulances
Road owners/managers	Finance consortia Private road managers	Local/national governments Public road managers
Industry	Producers of information technology hardware & software Automobile industries Road constructors Information technology linked sectors	
Interest organizations and public institutions		Motorist interest organizations Authorities Lobby organizations with strategic and local interests

Wrong problem

It is also noteworthy that information systems may sometimes be tackling the wrong problem. The need for real time information designed for a particular individual is contingent on the degree of reliability of the transport system. If the system were totally reliable, then the user could rely on standard timetables or a simple telephone call. The greater the unreliability of the transport system, the greater the need for real time information. In any case, sufficient efforts ought to be put into making the system more reliable rather than getting the user better informed.

Equity

All forms of technology have equity implications and the basic question here is whether transport telematics is socially cohesive or divisive. It may be argued that the car has resulted in greater inequality, as those having a car are benefiting through greater mobility and accessibility. Conversely, those without a car available have to rely upon increasingly poor public transport services affected by increased levels of congestion (cf. Masser et al. 1992). Their quality of life may deteriorate, as life styles become more car-oriented. With the new transport telematics technology, it is likely that these inequalities will increase, as those with access to the new technology benefit often at the expense of others. This polarization relates to the costs of the technology, but also to the knowledge and skill to use it. Society may be divided into the information literate and rich, and the information illiterate and poor. The question here is whether such a scenario is likely, and if so what can be done to reduce the level of technologically induced illiteracy and poverty. Interestingly, it may only be a transitory problem as the new younger generations are much more technologically literate than current generations.

To many people, the opportunities offered by transport telematics are very attractive, but the technologically led top-down approach may be tempered by the reality of some of the questions raised here. The transport telematics technology will not replace existing technology, and the two will exist in parallel for several years. In the short term, its greatest potential may be in systems about which the user has little knowledge (e.g. management systems) and in business applications where there are clear competitive advantages in using the technology.

Diffusion of innovation takes time, the technology itself often changes rapidly, and access to the technology is limited by knowledge and price. A major barrier to the acceptance of transport telematics is consumer awareness, the knowledge base required, and the realization of the full potential advantages that the technology will bring about. A more flexible approach to decision-making at the individual user or firm level is now possible, and this may in turn result in a greater variability in the performance of the transport system as a whole. Patterns of routinized behaviour, typical of many traditional transport studies, will be replaced by flexible dynamic choices which result from different decisions being made by people, often faced with the same choices. The challenge is not so much technological, but relate to the political and organizational issues, and to questions of social and economic acceptability.

3.6.3 Telematics Case Studies

From the description of the different types of telematics (Chapter 2) and the wide range of actors involved (Chapter 3), the potential for transport telematics is substantial; nevertheless, as we shall see its introduction has been limited. In the following five chapters, some of the successes and failures will be described, together with the lessons which can be learnt. The next chapter is concerned with the supply side of the debate in terms of the physical network. The potential for transport telematics is closely related to the availability and quality of the network as information is given to users on alternative routes and times. The chapter takes case study material from Italy, which is rather representative of a broad typology of the major European corridors. The supply side features are then in a subsequent part of the book (the Italian case study on motorway users) matched with the information needs through the use of Geographical Information Systems (GIS).

The second set of actors are the users of the system. There are chapters on car drivers

(Chapters 5 and 6), public transport users (Chapter 7), and road freight operators (Chapter 8). These are the three main actors involved, and together constitute the major demands for the infrastructure. In each case, empirical evidence is taken form a range of case studies in the Netherlands and the UK. Within the theoretical framework outlined in this chapter, the intention is to assess the nature and scale of use of the different forms of technology, both in terms of user response and in terms of the impact on the market as a whole. It is probably too early to assess whether the full market potential has been realized, but some answers can be given to the questions raised in Section 3.6.2.

The final two chapters address the broader questions of where the respective roles of the different actors should be concentrated. Chapter 9 includes a case study on managers of the road as they need to be convinced that through the use of transport telematics, a greater capacity will be obtained from the road network. Here the evidence on the attitudes and expectations of Dutch road managers is used to illustrate the choices to be made. It examines inter alia the questions of bottlenecks in institutions as well as on the infrastructure and the barriers to successful implementation. Chapter 10 includes recommended policy actions and actions from other public and private sector organizations.

Clearly, the performance of the transport system in Europe is not optimal, as its efficient operation is hampered by a wide range of external costs. One attraction of transport telematics is that reductions in the social costs of transport can take place without any increases in the prices charged. This option is acceptable to both politicians and the public. If it can be demonstrated that through the use of transport telematics, reductions take place in the levels of congestion, pollution and accidents, then there is likely to be widespread support for the technology. Subsequently it may be necessary to charge for the use of the technology, either directly to the users or indirectly through public agencies. Clearly, the acceptability of charging is likely to be much higher, provided that tangible benefits can be demonstrated and observed. Once this transition has been achieved, then the commercial potential of transport telematics can be fully realized (Banister 1995).

4 Potential Use of Telematics: A European Road Transport Corridor Typology

4.1 Introduction

The completion of the European market and the resulting opening up of the European space will induce a massive flow of people, goods and information all over Europe. Thus, Europe is increasingly moving towards a complex network economy, in which physical flows of people and goods form the connecting elements. This development has major implications for the transport system in Europe (see Banister et al. 1995; Button et al. 1996). Two major stress phenomena can be envisaged. First, metropolitan and urban areas will - as nodal points in the European networks - face severe economic and environmental externalities from transport. Second, the increasing spatial interaction between European nodes provokes major transport flows on all segments and corridors connecting these nodes. This means that European corridors - especially those between major metropolitan areas or centres of economic activity in Europe - are playing a critical role in the new evolving map of Europe. At the same time it has to be recognized that traffic congestion on many European transport corridors is dramatically increasing, while the environmental implications and the poor safety conditions are also causing unacceptable social costs. Conventional transport policy is apparently unable to keep up with Europe's growing demand for mobility.

It is clear that not all European transport corridors have the same characteristics. Therefore, it is necessary to develop an appropriate typology of European corridors on the basis of the indigenous relevant (e.g. physical, environmental, socio-economic) features of these corridors. The rationale behind the identification of corridor typologies is the idea that European policy resolutions (like the introduction of transport telematics) may generate different behavioural responses for each corridor type and its constituent links, so that it is necessary to have site-specific information in light of the fast dynamics on European transport arteries. Such a typology has the additional advantage that the results from given case studies (i.e. given road corridors and given telematics technologies) can *ceteris paribus* be transferred to other comparable situations (without claiming that all results have to be generalizable). The telematics technologies under consideration here are those applied in the field of inter-urban road transport. The main applications within this field are inter-urban traffic and road management, demand management, and inter-urban travel and traffic information.

The main goal of this chapter is to describe and analyze regions and road corridors/segments in Europe with regard to features that play an explanatory role regarding the relevance of transport telematics applications, including socio-economic, infrastructural and traffic demand related aspects. It should be emphasized that this goal does not imply an explicit assessment of the transport telematics market potential, but rather of the regional and corridor-specific conditions that influence this potential. Two case studies will be presented, each at a different spatial scale level. The first one (Section 4.2) focuses on European-wide corridors and their covered areas, for which a comparative analysis is carried out. The second one focuses on a local scale level; the area chosen here is Italy.

4.2 European Corridors: A Case Study

The present section intends to develop a practical typology of European inter-urban road corridors aiming at tracing the effects of possible telematics applications on constituent road links. In the case study presented, the identification of typologies is based on relevant attributes of a set of European transport corridors by employing all appropriate information on region-specific and site-specific links of these corridors in order to determine in a systematic way the potential of these connections for telematics introduction and application. Furthermore, ranking(s) of corridors according to the needs and potentials of telematics will be developed by means of multi-criteria analysis. Various results on European transport axes will also be shown[1].

4.2.1 Description of Relevant European Corridors and Their Features

In this section, we describe the specific European road corridors/segments used in our study including their regional and site-specific information. The first step is the selection of relevant inter-urban corridors (i.e. the identification of those inter-urban European road connections whose characteristics make them appro- priate for a possible implementation of telematics technologies). This issue depends inter alia on two aspects of the problem, viz. the *expected telematics impacts* (on systems efficiency, on network reliability, on environmental quality, etc.), and the *modus operandi of different actors* in the transport system concerned (i.e. users, operators, public decision-makers and private decision-makers in the economic and social system as described in Chapter 3). From the large variety of different inter-urban road corridors in Europe, a selection based on plausible criteria has been made on the basis of the following criteria:

- the *geographical site and spatial coverage* of a corridor; this criterion refers to origin-point destinations and to classes of regions being connected (e.g. central and peripheral areas, the strategic importance of the corridor, connectivity with

[1]The material presented here is largely based on the work carried out in MARTA (1993c).

metropolitan areas etc.).
- the *site-specific mobility patterns* on these corridors; such aspects relate to congestion degrees in certain segments of the corridor (e.g. in the vicinity of large urban centres), environmental pollution in given areas, network reliability and safety on certain segments, etc. Such aspects will in our study be considered in relation to region-specific socio-economic conditions (e.g. in terms of welfare, sectoral specialization etc.).

Based on the above considerations, an identification of various important and representative European corridors which lend themselves also for interesting scientific analysis has taken place, with a cross reference to a number of projects in the European DRIVE programme which concerned inter-urban telematics applications (EC 1993). These projects included the key corridors in Europe with potentials for telematics applications. The selected projects were PLEIADES (Paris-London Corridor), METAFORA (Major European Testing of Actual Freight Operations Using RTI on an Axis), MELYSSA (Mediterranean-Lyon-Stuttgart Site for ATT), ADS (Automatic Debiting Systems), INVAID II (Evaluation of the INVAID Automatic Incident Detection System), COMBICOM (Combined Transport Communication Systems) and FRAME (Freight Management in Europe), which all together cover Centre-East and Centre-West directions in Europe. In order to consider a more balanced mobility and flow pattern, an additional set of five corridors is taken into account reflecting East-West directions (such as Trieste-Lisbon and Vienna-Nantes corridors) as well as inner links in the core of Europe (such as London-Berlin, Amsterdam-Milan and Amsterdam-Utrecht) (see also Figure 4.1).

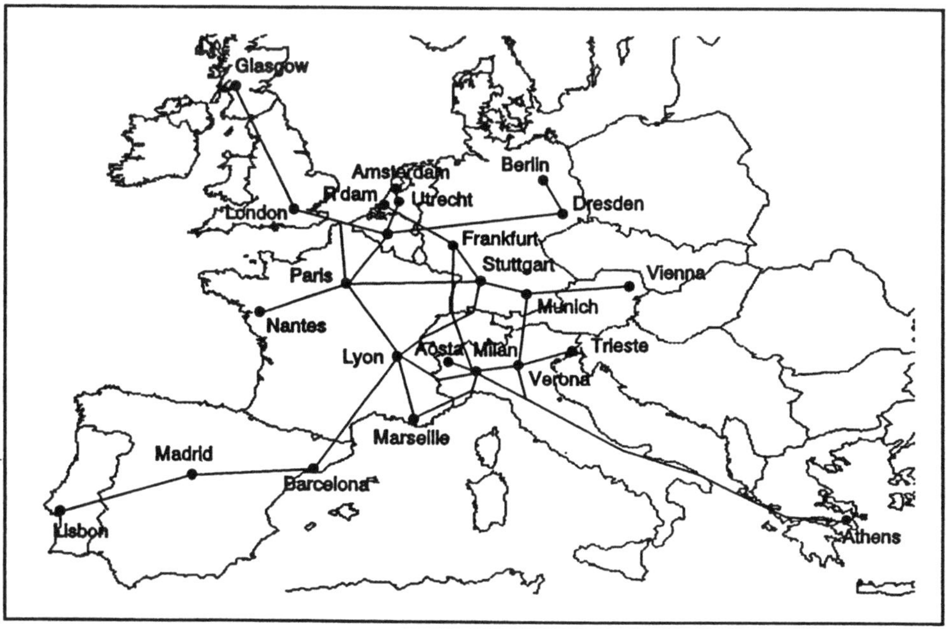

Fig. 4.1. European inter-urban corridors under study.

Consequently, after an exploratory analysis on the basis of the above criteria a list of corridors has been identified (Table 4.1). Some of these corridors may be peripheral in nature, others central, or of a mixed nature. Each corridor route is identified on the underlying basis of the main road infrastructure (motorway, or highway). Figure 4.1 also shows that corridors are subdivided into shorter segments between major cities connecting two relevant nodes along the route; this disaggregation can be fruitfully utilized in the subsequent traffic indicator analysis. It is noteworthy that the selected twelve corridors reflect the main characteristics of the European transport network:
- a *larger number of corridors* in North-South direction than in East-West direction;
- a *denser network of corridors* in the European core area in comparison with the infrastructural links between centre and periphery.

Table 4.1. European inter-urban road corridors under study.

Corridor	Used abbreviation	Corridor	Used abbreviation
1. Glasgow-London-Paris-Marseille (PLEIADES)	Gla-Mar	6. Rotterdam-Munich-Verona (COMBICOM)	Rott-Ver
2. London-Athens (METAFORA)	Lon-Athe	7. Amsterdam-Athens (FRAME)	Ams-Athe
3. Stuttgart-Lyon-Barcelona-Madrid (MELYSSA)	Stu-Mad	8. Trieste-Lisbon	Tri-Lis
		9. Vienna-Nantes	Vie-Nan
4. Aosta-Athens (ADS)	Aos-Athe	10. London-Berlin	Lon-Ber
5. Amsterdam-Madrid (INVAID II)	Ams-Mad	11. Amsterdam-Milan	Ams-Mil
		12. Amsterdam-Utrecht	Ams-Utr

It should be emphasized that - although the focus is on long distance corridors - all these corridors are subdivided into links or segments which are also individually considered. This means that the level of analysis concerns both the corridor as a whole and its constituent links, as is also evident from the underlying region-specific socio-economic data and the site-specific traffic data.

In order to study the differences in the potential telematics impacts as consequences of different segment features (leading to a consistent corridor typology), a first stage is the construction of the 'states' of the corridors. This requires the identification of the main categories of criteria necessary to measure the characteristics of an inter-urban corridor and its constituent links consistent with the objective of assessing the telematics potential. In particular, we have taken into account the effect on:

- *the mobility systems' efficiency*, particularly with regard to management costs and use by operators, and to new network services and capacities, etc.;
- *the network reliability*, with regard to higher safety, and better information and service levels for users;
- *the environmental quality*, particularly with regard to pollution reduction.

Based on these elements, a typology of states of corridors can be derived, by using the following qualitative characteristic features:
- *socio-economic conditions* (within the regions crossed) which are determining the different mobility patterns and, as a consequence, the different needs in terms of

telematics functions as well as the different behavioural responses;

- *strategic network position* (in other words, the relative locational advantage of the corridor) which is useful to capture the importance and the effective role of the corridor concerned, as well as to provide indirect (quantitative or qualitative) information on supply-side impacts of telematics. Despite the fact that all selected corridors may be seen as strategic from a social and economic perspective, their different relative position - which is region and site-specific - differentiates the importance of this characteristic feature.
- *institutional network organization*, which deals inter alia with the managerial network perspective and refers also to the level of the infrastructure supply as well as to the degree of complexity of management (e.g. the degree of interrelation/cooperation among transport operators along the corridors, the different complexities of infrastructural management, the variety of managers in the infrastructural administration, the different operative strategies adopted and, finally, the different levels of technological innovation already achieved).
- *site-specific traffic conditions*, with particular reference to: (i) the degree of congestion (i.e. the identification of bottlenecks on links along the corridors which are generated by an insufficient supply of infrastructure (or by an excess demand) and (ii) safety (reliability or accident rates on links on the corridor), which may influence the share of social (direct or indirect) benefits as a result of the introduction of transport telematics;
- *environmental quality* (noise annoyance or air pollution), which is one of the principal motives for implementing transport telematics.

Some tentative indications on the association between 'states' of the corridors and the potential impacts of transport telematics are provided in Table 4.2 (for more details on the various kinds of potential impacts of transport telematics we refer to Nijkamp et al. 1992). They are presented here as plausible impact signs in summary form. The following concise observations can be made on the successive elements of this table.

- The differences in *socio-economic situations* on corridors and their constituent links can be identified by the regions crossed with respect to the 'core' of Europe (Section 4.2). Moving from the centre towards the periphery, the residential density as well as the per capita value added decreases, while production activities become increasingly industrial-oriented. Consequently, it is plausible to argue that on corridors in the central regions, the principal aims of telematics technologies are to provide traffic information and route guidance, driver support, automatic debiting, etc., which will have a high socio-economic benefit (both in terms of market potential and behavioural responses), while in the peripheral regions, in which the share of heavy vehicles is relatively high, telematics equipment in the field of freight and fleet management may be more relevant.
- The *strategic network position* shows a large positive influence on telematics introduction, particularly regarding direct effects on the users, but also for indirect effects (social and organizational efficiency).
- By considering the state of the *institutional network organization*, it seems that a maximum telematics potential will result in a medium to high level of management complexity. However, the potential effects on specific indicators are likely to be

different. The greatest influence would be on organizational structure, with lesser influence on direct and indirect user benefits, and only marginal influence on locational shifts.
- Among the *traffic conditions* on links and the corridor as a whole, we have to look in particular at congestion. Even though there is in general, a negative relationship between overall levels of efficiency and the degree of congestion, a maximum gain in efficiency (deriving from telematics) is supposed to be reached for a medium to high degree of congestion. In this situation, telematics technologies have the greatest potential to influence behaviour. In a situation of low or very high congestion, the telematics impact will be marginal.
- The last two states (*safety and environmental quality*) have various effects on socio-economic indicators; the former directly affects users, the latter has indirect social effects. The influence on organizational and locational effects is marginal. Probably the telematics impact on safety is high when reliability is low. The impact on environmental quality is of course relevant for the reduction of pollution.

Table 4.2. Plausible relationships between the 'state' of the corridors and the potential impacts of transport telematics.

Socio-economic conditions			Locational conditions			Institutional network organization			Traffic condition: congestion			Traffic condition: safety			Environmental quality		
L	M	H	C	S	P	L	M	H	L	M	H	L	M	H	L	M	H
++ +	+ +	+	+	+ +	++ +	0	++ +	+	0	++ +	+	++ +	+ +	+	++ +	+ +	+

Legend: L, M, H = low, medium and very favourable states of the corridors, respectively; C, SP, P = central, semi-peripheral and peripheral locational conditions, respectively; 0 = impact negligible ++++++ = increasing impact levels

Our data have been collected based on the above considerations. In total, 19 different features have been considered. As mentioned before, they are to be regarded as factors influencing impacts of telematics in terms of user benefits, efficiency, safety and environmental quality. The classification of these features is based on criteria related to socio-economic conditions (1), strategic network position (2), road infrastructure and institutions (3), road traffic demand (4), and environmental impacts of road traffic (5).

1. *Socio-economic* features concern characteristics of the regions located around the corridors. Criteria related to social conditions include population densities and stocks of dwellings. Criteria describing economic activities in these areas include productivity and employment figures. The spatial differences in economic patterns are relevant for an assessment of structural mobility patterns between economic core regions. These factors determine the economic importance of corridors between respective core regions and consequently the importance of advanced traffic management tools on these corridors. Furthermore, an assessment of the size of the market for both public and private inter-urban telematics applications will be partially based on direct market variables, such as income structures in various regions. These may influence the ability and willingness to invest in road-based telematics systems or to purchase private consumer

applications. The assessment of these various characteristics for each corridor investigated is done by using mean figures of the set of its surrounding regions. Such an assessment is based on the standard regional subdivision of Western Europe into so-called NUTS I regions[2].

2. Criteria related to *strategic locations* on the corridors include the large agglomerations on the corridors and the number of different Community member states that are crossed. The former feature refers to the economic strategic importance of corridors, the latter to the complexity of the mix of the different (national) policies towards road infrastructure that govern the same corridor.

3. Criteria related to *road infrastructure* include motorway densities in the surrounding areas of the selected corridors and the accessibility of important agglomerations connected by the corridors. These variables are an indication of possible alternative routes between various links. In this respect, motorway densities are a valid proxy for short-distance alternative routes (relevant in the case of dynamic route guidance systems), while the latter provides insight into the quality of infrastructure at a broader European level.

4. Criteria related to *road traffic demand* include private car ownership, road traffic accident rates, vehicle flows and related growth, flows of heavy vehicles and congestion levels on the corridors. Private car ownership is an indicator for the car traffic generating potential in the various considered regions. The latter five features concern the characteristics related to motorways themselves. The number of road traffic accidents is a measure for safety conditions. High congestion levels determine the level of externalities of traffic and thus indirectly the potential benefit of telematics for traffic efficiency and environmental quality. It is also evident that corridors with large traffic flows have a higher potential for telematics implementation than corridors with lower traffic volumes, and that corridors which form part of important freight transport axes require telematics applications that serve to improve specific problems related to freight operations (e.g. hazardous goods monitoring).

5. As a relevant proxy for the *environmental* criterion, the level of NOx emission has been taken. The general observation that on average more than 60% of total NOx emissions in Europe is caused by car traffic makes this feature a valid proxy for the environmental implications of road traffic in regions crossed by the selected corridors.

In Table 4.3 a tentative assessment is given of how these specific features may influence the corridor-specific potentials of transport telematics.

A summary of the characteristics of the data including their source, year and definition is shown in Table 4.4. Most of these indicators concern measurable variables. Many of them are taken from Eurostat Regional Statistics, in which extensive definitions can be found. The indicators 'accessibility index' and 'congestion level' (LOS) need further explanation. The accessibility indices for European regions in our study are taken from Keeble et al. (1982), where the measurement of the accessibility of European zones is given by using a simple gravity type formulation in which distance is the main indicator:

$$A_i = \Sigma_j M_j/d_{ij}^c$$

---

[2]The level I of the Eurostat NUTS (Nomenclature of Territorial Units for Statistics) regional subdivision of Europe consists of 72 European Community Regions (RCE's) (Eurostat, 1991).

where
A_i = accessibility of zone i
M_j = regional gross domestic product in ECU
d_{ij} = shortest distance in Km between centre of zones i and j
c = impedence parameter.

Table 4.3. Factors causing spatial variations in the inter-urban potential of transport telematics.

Factor:	Influence on:
Socio-economic structures	
economic spatial and sectoral compositions	urban network interactions
income levels	road authorities telematics investment budgets consumer purchase power
Strategic location	
strategic location	policy driven telematics potential
Infrastructural	
motorway density	use of telematics information for alternating routes
Road traffic demand	
car ownership	potential private telematics consumer market
congestion levels	potential contribution of telematics to improve traffic performance
traffic flow	potential private telematics consumer market potential reduction in externalities
freight traffic flow	importance of freight transport monitoring and control
accident rates	safety benefits of telematics
Environmental	
traffic pollution	telematics benefits to environment

This is clearly a simple but practical approach (see for more details on accessibility studies, Martellato et al. 1996). In our case study the gravity parameter c is assumed to equal 1. The road accessibility index for a given corridor is the average of the indices of the respective zones crossed by the corridor.

The 'Level Of Service' (LOS) index relates road capacities to (measured) actual traffic volumes. Therefore, it gives an approximation for the smoothness of the traffic situation on a given road segment, and it implies an estimation of the road capacities of European highways. While for some American case studies the (theoretical) road capacities have been assessed by field observation, in Europe no structural research has been carried out so far on thi subject. The main concept used here is the existence of an optimal average speed (the so-called 'critical speed'). At this speed, the theoretical capacity of the road is at its maximum. This theoretical capacity is dependent on various factors, related to the characteristics of vehicles using the road and the road infrastructure itself (Nijkamp et al. 1993). Next this capacity level is compared with the actual performance. In this way, a reasonable and realistic proxy indicator can be derived. All data from Table 4.4 have been collected for the above mentioned transport telematics-relevant segments and corridors in Europe.

Table 4.4. List of features for the corridor typology with source, year and measurement unit.

Nr.	Variable	Source	Year	Measurement unit
	Socio-economic related			
1	population density	Eurostat	1985	habitants per km2
2	(Total) GVAMP (*)	Eurostat	1984	ECU per habitant
3	GVAMP of industrial sectors	Eurostat	1984	ECU per habitant
4	GVAMP of market services	Eurostat	1984	ECU per habitant
5	employment	Eurostat	1984	workforce per km2
6	employment in industrial branches	Eurostat	1984	workforce per km2
7	employment in market service branches	Eurostat	1984	workforce per km2
8	stock of dwellings	Eurostat	1985	dwellings per km2
	Locational related			
9	large agglomerations on corridor	-	1992	number of large agglomerations on corridor
10	crossed EC member states	-	1992	number of national borders crossed by corridor
	Infrastructure related			
11	motorway density	Eurostat	1985	length of motorway (km) per km2
12	accessibility of crossed zones	Keeble	1982	accessibility index; see explanation
	Traffic demand related			
13	private car ownership	Eurostat	1985	private owned cars per km2
14	road traffic accidents	Eurostat	1985	motor vehicle traffic accidents per km2
15	vehicle flows	NEA	1985	passing vehicles in yearly daily averages
16	growth of vehicle flows	NEA	1985	relative growth of vehicle flows between 1980 and 1985
17	heavy vehicle flows	NEA	1985	share of heavy vehicles
18	congestion levels	NEA	1985	LOS index; see explanation
	Environment related			
19	NOx pollution	Europe 2000 (CEC)	1985	NOx pollution in tons per km2 per year

(*) GVAMP = gross value added at market prices.

4.2.2 Comparison of Corridors from the Viewpoint of Their Potential Telematics Applications

As shown in the previous section, the selected site-specific attributes of corridors can provide relevant (qualitative) information on the potential impact of transport telematics applications (Reggiani et al. 1993 and 1995). In this section the main differences and similarities in these corridors will be examined in order to identify major characteristics of the selected inter-urban corridors with a view on their suitability for specific classes of transport telematics.

From the point of view of *'socio-economic' attributes* (Figures 4.2 and 4.3), a subdivision into *central* and *peripheral* European regions comes immediately to the fore. In particular, it is noteworthy that by moving from the European peripheral regions towards the central ones (in other words, moving from the corridors connecting centre with periphery towards the corridors and links joining central nodes in Europe), the residential density increases as well as the share of the service in total employment. In addition, by considering only the centre-periphery corridors, differences in residential density along their constituent links become apparent, since those in a North-South direction have an intermediate level of residential density compared to those in an East-West direction.

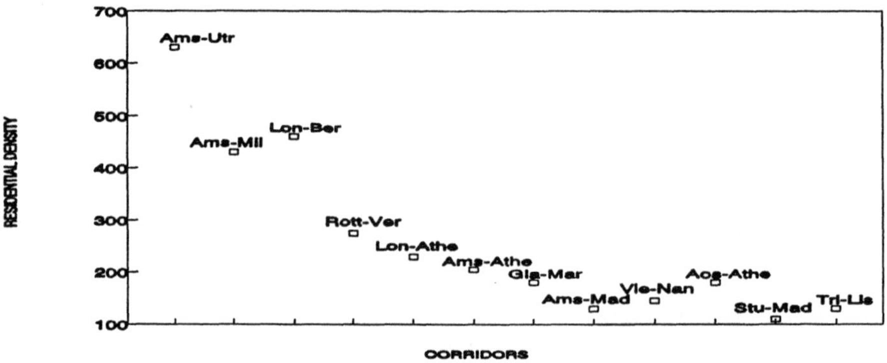

Fig. 4.2. Residential density (1987; in habitants per km2) in the corridor areas.

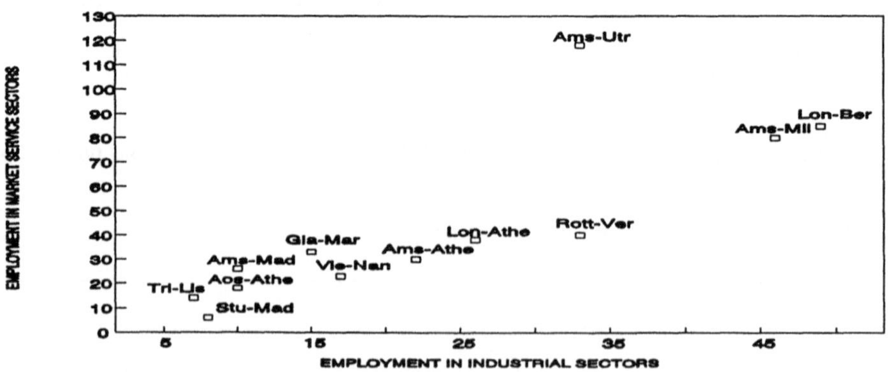

Fig. 4.3. Employment in industrial sectors versus employment in service sectors in corridor areas (1987; in habitants/km2).

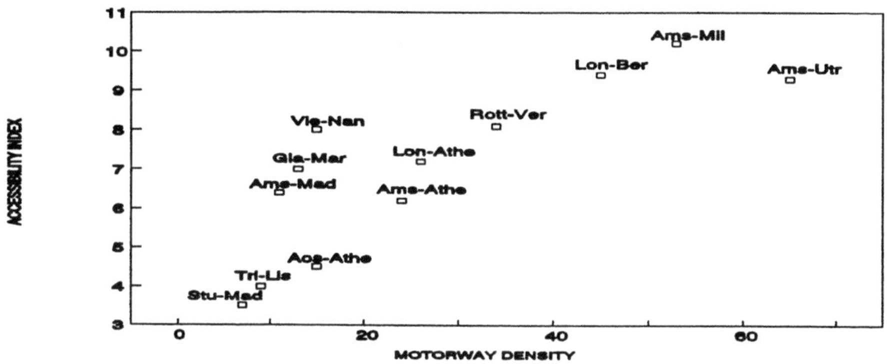

Fig. 4.4. Motorway density (1985; in km/1000 km2) versus the accessibility of the corridor areas (1988).

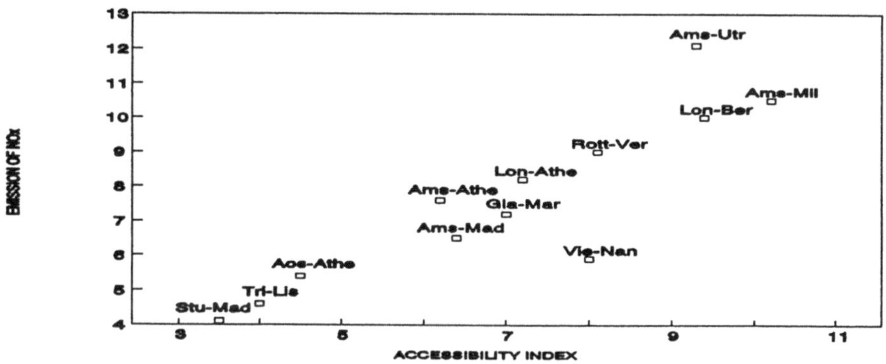

Fig. 4.5. Accessibility (1988) versus emission of NOx (1985; in tons/km2/year) in the corridor areas.

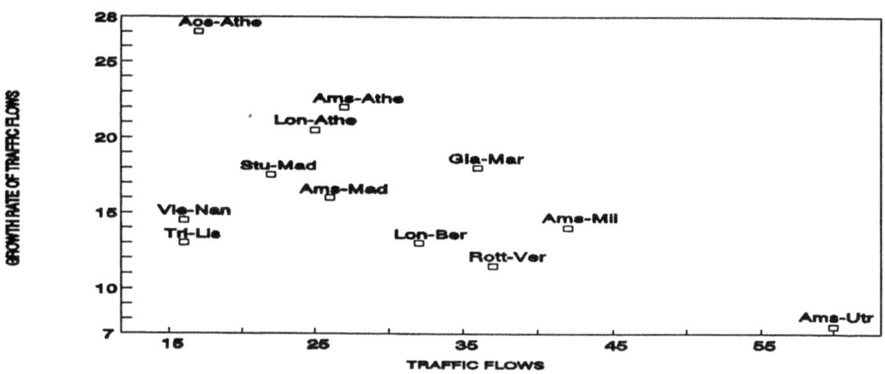

Fig. 4.6. Traffic flows (1985; in 1000 veh./day) versus growth rate of traffic flows (between 1980 and 1985; in %) on the corridors.

As a consequence, from these site-specific differences in socio-economic attributes it is plausible that different needs will emerge, both in terms of mobility and of specific telematics implementations, as well as different reactions of potential users. The differences in mobility (both in terms of the functioning of the system and the user needs) are confirmed by considering the *infrastructure supply side* (Figure 4.4). In particular, this figure shows, on the basis of data on motorway density, the existence of an interesting global classification, namely the central corridors and peripheral corridors, while on the basis of an indicator of accessibility (which reflects the different effects of dominance and dependence among the regions) a further subdivision into *quasi-peripheral* and *peripheral* corridors can be made. This is mirrored by a high, medium and low supply of infrastructure, respectively.

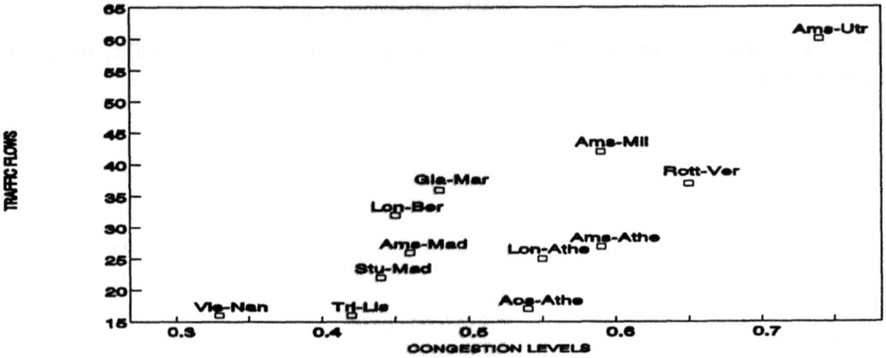

Fig. 4.7. Congestion levels (1985) versus traffic flows (1985; in 1000 veh./day) on corridors.

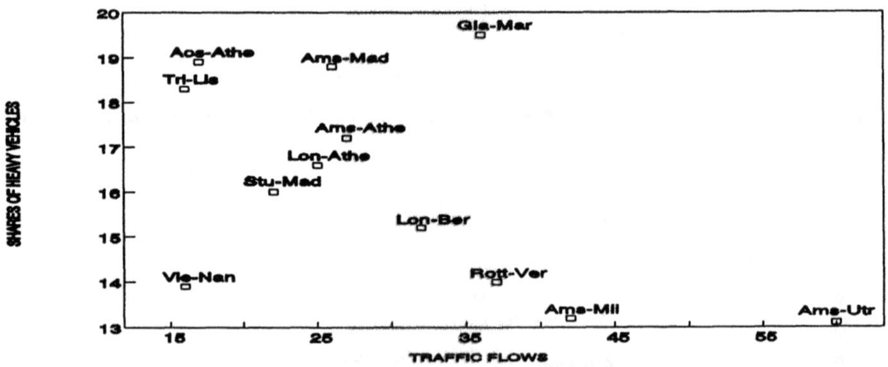

Fig. 4.8. Traffic flows (1985; in 1000 veh./day) versus the share of heavy vehicles in the traffic flow (1985; in %) on the corridors.

Thus, by combining infrastructure supply and socio-economic attributes, a classification emerges (i.e. central, peripheral and quasi-peripheral corridors), which is mainly based on the geographical position of the corridors with respect to the core of Europe.

With respect to the *transport demand attributes*, it is possible to identify the same type of classification which mainly reflects average congestion situations on European links. In particular, by investigating Figures 4.5 and 4.6 we can see that on average the corridors connecting peripheral areas are characterized by a low accessibility index in combination with a low level of air pollution, as well as a low value of traffic flows. It is noteworthy from Figure 4.6 that the traffic growth on routes with a lower traffic flow is rather high, which suggests that - as a result ofincreasing trade - all European corridors tend to move to problematic traffic situations. Next, by considering the average level of services related to traffic flows and the share of freight transport on total flows (Figures 4.7 and 4.8, respectively) again the peripheral corridors appear to be clustered and clearly separated from the other central corridors.

In conclusion, on the basis of the above diagrams reflecting quantitative information on socio-economic, transport supply and transport demand attributes, two main categories of European corridors emerge. First, there is a category of corridors which connect *central* areas (Amsterdam-Utrecht, Amsterdam-Milan, Rotterdam-Verona, London-Berlin, London-Athens and Amsterdam-Athens). These corridors and segments are characterized by a high infrastructural density and degree of accessibility, but are clearly insufficient in satisfying the total traffic flows (see the high congestion level and pollution emission). Second, there is a category of *(quasi) peripheral* corridors which connect more isolated areas in Europe (Aosta-Athens, Glasgow-Marseille, Amsterdam-Madrid, Stuttgart-Madrid, Trieste-Lisbon and Vienna-Nantes). These connections are on average characterized by a medium congestion level and a low accessibility and show the necessity for infrastructural investment in order to avoid a rapid increase of congestion.

It should be noted that the above classification is based on averaging the data of all links pertaining to the same corridor. Then the expression (Quasi)peripheral corridors denotes that - in these corridors - the peripheral links are strategically more important than the core links. As a consequence, the above corridors are typified by their corresponding periphery, while the 'central corridors' are characterized by their corresponding relevant core. This observation also leads to the need for differentiating telematics technology - for the same corridors - according to peripheral and core characteristics. Clearly, a further detailed analysis can be only given by using a 'nested' typology within the above mentioned corridors.

Finally, it should be noted that the above two-fold classification can be confirmed by means of a further analysis of the interdependencies among all relevant road attributes, by using a multi-criteria approach. This analysis will be discussed in Section 4.2.3. However, it is already evident from our first results that for each of the two groups of corridors, significant differences in both the potential of transport telematics applications and the related behavioural responses are likely to emerge.

4.2.3 A Classification of European Inter-urban Corridors by Means of Multicriteria Analysis

Methodology

In the previous section we have offered a preliminary global typology of European corridors on the basis of regional/local data, which has led to a clustering of the corridors into two main classes, namely central and peripheral corridors. It should be noted that the data on socio-economic, supply and demand features constitute a multi-dimensional profile of corridors which presents different aspects of the 'socio-economic traffic' profile of the corridors concerned in relation to the potential introduction of telematics. A more thorough evaluation of this potential may then be based on a multi-criteria analysis (MCA), in order to design a ranking scheme of corridors in relation to their 'compound value', determined by the interdependencies of the various attributes of the links (Table 4.5).

MCA has become a popular tool in policy evaluation studies in many countries (e.g. Nijkamp 1979 and Nijkamp et al. 1990), not only in the case of 'hard' (quantitative) information but also in the case of 'soft' (qualitative) information. In this framework, a suitable MCA method, namely the 'Regime Method', which meets the criteria of methodological soundness, mathematical and statistical accessibility and easy computer use, can be used (Hinloopen et al. 1983 and Hinloopen and Nijkamp 1990). This method is also used in the present study.

Table 4.5. Criteria used in the MCA analysis.

a.	**Socio-economic profile**	**c.**	**Transport demand profile**
a1	Population density	c1	Private car ownership
a2	(Total) GVAMP (*)	c2	Road traffic accidents
a3	GVAMP of industrial sectors	c3	NOx pollution
a4	GVAMP of market services	c4	Vehicles flows
a5	Employment	c5	Growth of vehicle flows
a6	Industrial employment	c6	Heavy vehicle flows
a7	Service employment	c7	Traffic flow smoothness
a8	Stock of dwellings		
b.	**Transport supply profile**		
b1	Motorway density		
b2	Accessibility of cities		
b3	Large agglomerations on corridor		
b4	Crossed EC member states		

(*) GVAMP = Gross value added at market prices

Analogous to Section 4.2.2, we distinguish again three main evaluation criteria, namely the Socio-Economic Profile, the Transport Supply Profile and the Transport Demand Profile. Each of these main judgment profiles is composed of observable attributes, based on the data collected as shown in Table 4.5. Here, 19 indicators, grouped into three main criteria categories, are used. The resulting so-called impact matrix is shown in Table 4.6 as a set of ordinal rankings of all routes for each relevant attribute. The typology based on an MCA approach may use equal or different weights for each of the main criteria and sub-criteria, depending on actor priorities. Five sets of ordinal rankings of importance attached to the various critera have been used in our study by way of sensitivity analysis (Table 4.7).

Table 4.7 shows for each of these five weighting scenarios an ordinal ranking of attributes according to their expected relevance. In scenarios I and II, priority is given to the attributes related to the 'traffic' criteria (both supply and demand) and to 'socio-economic' criteria, respectively. In scenarios III and IV the 'socio-economic' criteria are not taken into account, so that corridors are only classified in the context of transport characteristics. In Scenario III more relevance is then given to 'demand-side' attributes, and in Scenario IV to 'supply-side' attributes. Finally, in Scenario V a high preference is only given to socio-economic data. For each scenario a Regime analysis is carried out. The results are presented in the next section.

Table 4.6. The impact matrix (complete rankings).

	Socio-economic profile								Transport supply profile			
Corridor	a1	a2	a3	a4	a5	a6	a7	a8	b1	b2	b3	b4
Amsterdam-Utrecht	12	8	6	11	12	10	12	12	12	10	1	1
Amsterdam-Milan	11	12	11	12	10	11	10	11	11	12	6	10
London-Berlin	10	9	8	8	11	12	11	10	10	11	2	9
Rotterdam-Verona	9	10	12	9	9	9	9	9	9	9	5	6
Amsterdam-Athens	7	6	9	5	7	7	6	7	7	4	11	11
London-Athens	8	5	7	6	8	8	8	8	8	7	12	12
Glasgow-Marseille	6	7	5	7	6	5	7	6	4	6	8	2
Amsterdam-Madrid	3	4	3	4	3	4	4	3	3	5	9	5
Aosta-Athens	5	1	1	1	4	3	3	4	5	3	3	3
Trieste-Lisbon	2	2	2	3	2	2	2	2	2	2	10	7
Stuttgart-Madrid	1	3	4	2	1	1	1	1	1	1	4	8
Vienna-Nantes	4	11	10	10	5	6	5	5	6	8	7	4

	Transport demand profile						
	c1	c2	c3	c4	c5	c6	c7
Amsterdam-Utrecht	12	12	12	12	1	2	12
Amsterdam-Milan	11	10	10	11	3	1	10
London-Berlin	10	11	11	8	4	5	4
Rotterdam-Verona	9	4	9	10	2	4	11
Amsterdam-Athens	7	6	7	6	11	8	9
London-Athens	8	9	8	5	10	7	8
Glasgow-Marseille	6	7	6	9	9	11	6
Amsterdam-Madrid	3	5	4	7	7	10	5
Aosta-Athens	4	8	3	3	12	12	7
Trieste-Lisbon	2	3	2	2	5	9	2
Stuttgart-Madrid	1	1	1	4	8	6	3
Vienna-Nantes	5	2	5	1	6	3	1

Results

The results of our MCA (by applying a Regime method) within the framework of the 'complex' evaluation of twelve European inter-urban corridors for telematics introduction are presented in Table 4.8. Each of the five columns in this Table refers to the corresponding Scenarios I...V described in Table 4.7. Two types of results emerge from Table 4.8. First, from these results we can draw the conclusion that the two-fold classification derived in Section 4.2.2 seems fairly robust against changes in the various scenarios as tested by an MCA. This is reflected by the position of corridors in the upper and lower blocks in Table 4.8. In other words, the Multicriteria Analysis displays the highest scores for the central group of the corridors in light of all 'socio-economic traffic' features. Thus, a clear subdivision between central and quasi-peripheral corridors emerges again. It should be noted that the rationale behind the definition of such a classification is the hypothesis that differences in the global average telematics impact on corridors depends on hierarchical levels based on 'socio-economic traffic' characteristics.

Table 4.7. Ordinal rankings (in decreasing order of priority) of the attributes in five weighting sets (representing five scenarios) I, II, III, IV and V.

SCENARIO I	c7	c4	c5	c6	b3	b4	c3	b1	c2	b2	c1	a1	a8	a2	a5	a3	a4	a6	a7
SCENARIO II	a7	a6	a4	a3	a5	a2	a8	a1	c1	b2	c2	b1	c3	b4	b3	c6	c5	c4	c7
SCENARIO III	c4	c3	c7	c5	c6	c2	c1	b2	b4	b3	b1								
SCENARIO IV	b1	b3	b4	b2	c1	c2	c6	c5	c7	c3	c4								
SCENARIO V	a1	a8	a2	a5	a3	a4	a6	a7											

Table 4.8. Five-cluster typology of corridors as a result of multicriteria analysis.

Scenarios					
I		II		III	
i	Amsterdam-Utrecht	i	Amsterdam-Utrecht	i	Amsterdam-Utrecht
i	Amsterdam-Milan	i	Amsterdam-Milan	i	Amsterdam-Milan
i	London-Berlin	i	London-Berlin	i	London-Berlin
i	Rotterdam-Verona	i	Rotterdam-Verona	i	Rotterdam-Verona
ii	London-Athens	ii	London-Athens	ii	London-Athens
ii	Amsterdam-Athens	ii	Amsterdam-Athens	ii	Amsterdam-Athens
iii	Glasgow-Marseille	iii	Glasgow-Marseille	iii	Glasgow-Marseille
iv	Aosta-Athens	v	Wien-Nantes	iv	Aosta-Athens
iii	Amsterdam-Madrid	iii	Amsterdam-Madrid	iii	Amsterdam-Madrid
v	Wien-Nantes	iv	Aosta-Athens	v	Trieste-Lisbon
v	Trieste-Lisbon	v	Trieste-Lisbon	v	Stuttgart-Madrid
v	Stuttgart-Madrid	v	Stuttgart-Madrid	v	Wien-Nantes
IV		V			
i	Amsterdam-Milan	i	Amsterdam-Utrecht		
i	London-Berlin	i	Amsterdam-Milan		
i	Amsterdam-Utrecht	i	London-Berlin		
ii	London-Athens	i	Rotterdam-Verona		
ii	Amsterdam-Athens	ii	London-Athens		
i	Rotterdam-Verona	ii	Amsterdam-Athens		
iii	Glasgow-Marseille	iii	Glasgow-Marseille		
iv	Aosta-Athens	v	Wien-Nantes		
iii	Amsterdam-Madrid	iv	Aosta-Athens		
v	Wien-Nantes	iii	Amsterdam-Madrid		
v	Trieste-Lisbon	v	Trieste-Lisbon		
v	Stuttgart-Madrid	v	Stuttgart-Madrid		

Secondly, the above distribution of corridors into the upper and lower blocks of Table 4.8 can be further analyzed by considering not only their position relative to the core of Europe, but also by investigating more precisely their geographical orientation with respect to the core area. By combining these dimensions, a *five-cluster typology* of inter-urban corridors, which show in particular similar profiles with respect to their transport demand attributes, emerges, namely the centrally located corridors (i), semi-peripheral, eastward-oriented corridors (ii), semi-peripheral, westward-oriented corridors (iii), peripheral, eastward-oriented corridors (iv), and peripheral, westward-oriented corridors (v). The various corridors in Table 4.8 are characterized by these 5 features (i) - (v). It is interesting to note that the results in Table 4.8 display - for each scenario - a descending hierarchy in terms of eastward- and westward-oriented corridors from the core of Europe, which suggests the necessity to differentiate also in these directions.

The previous results are of a global and average nature. Our data allow also more region- and site-specific analysis. Thus also various individual segments can be considered. Such detailed site-specific information generates an interesting mapping of clusters of problem situations on European road corridors, with a particular focus on problem cases on the routes in the European core area. Combination of all information in a qualitative sense leads to a set of contour lines, starting from the core of Europe with a descending intensity of problem cases on road segments. The spatial shape suggests some sort of an egg form, which is somewhat similar to the European 'blue banana'. The combined mapping of these clusters, showing similar transport demand profiles, appears to result thus in a set of concentric zones, called the *Euregg* Model (Figure 4.9).

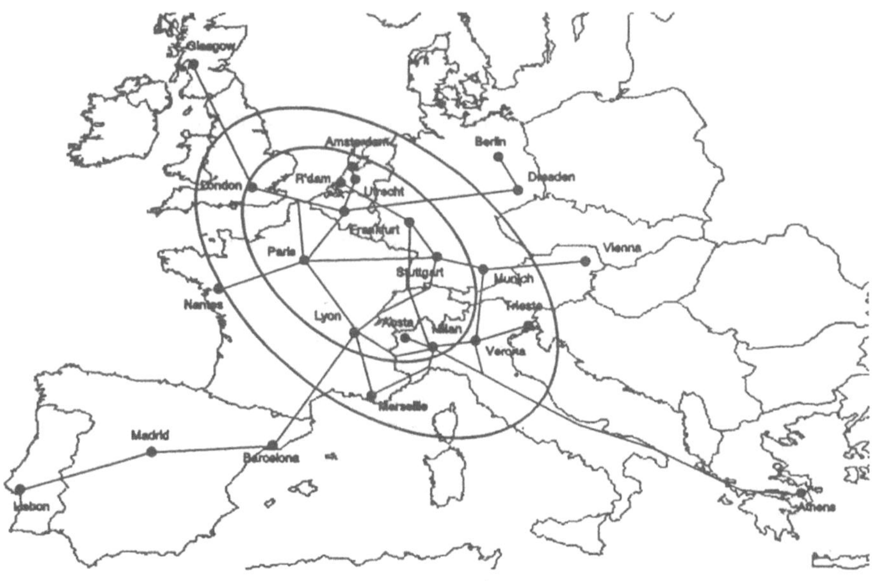

Fig. 4.9. The design of an inter-urban corridor typology in five clusters: the Euregg Model.

4.2.4 Conclusions

Externalities (congestion, fatalities, environmental pollution) create some of the most severe problems in European transport systems and road corridors. Such externalities are not only confined to the road users themselves, but have also high social costs for society at large. If expansion of infrastructure is not a viable option, a better management of the existing capacity is necessary. In this framework, telematics may play a critical role.

At the same time it has to be recognized that the provision of up-to-date and real time information may be helpful in improving the inefficiency inherent in traffic on many corridors, but that the ultimate benefits will depend on user responses to this new technology (or a set of distinct new technologies). Furthermore, the implementation and operation of telematics systems have to be financially viable, and therefore it is important to gain sufficient insight into the future market potential and penetration of such systems on various classes of European transport arteries. Furthermore, it is necessary to pay due attention to behavioural responses of latent road users, as improvement of the quality of a corridor may attract new traffic. Therefore, telematics needs sufficient behavioural research on site-specific conditions and responses.

This case study has tried to develop a systematic methodology which has been tested for European road corridors in order to derive a practical typological approach to the telematics market. Various empirical results were obtained. A full listing of European road segments from the viewpoint of problem cases and telematics potentials may help in building a more appropriate prioritization scheme for transport telematics investments in Europe. It is clear that a next step would be to identify which specific telematics technologies are needed to cope with site-specific problem situations on European routes. This will require, as the next step of a meaningful research agenda, the development of a transport telematics technology assessment methodology.

4.3 Regional Corridors: A Case Study of Italy

A comparative evaluation of long transnational corridors with respect to the potential of transport telematics is a difficult task, since across such test areas there is no spatial homogeneity of telematics-relevant characteristics: problem areas are often those around main urban agglomerations or may be the result of physical barriers and bottlenecks in the motorway network. Looking at the spatial scale level of regions and local segments of motorways that have uniform characteristics - rather than entire motorways - a more detailed analysis can be carried out. The chosen case area is Italy, a country with significant spatial contrasts between North and South in terms of socio-economic profile[3].

The analysis has a descriptive character. It is based on an assessment of mobility needs at the inter-urban level in Italy. It takes into account both traffic conditions and structural characteristics of the urban and regional system, which are basically social and economic in nature. The analysis is supported by the use of Geographical Information Systems (GIS), which allows us to visualize various types of spatial data related to local motorway segments. GIS-based digital maps of roads are able to assist us to identify segment-specific local problem cases.

---

[3]The material presented here is largely based on the work carried out in MARTA (1994c).

4.3.1 Characteristics of Urban and Regional Networks

A marked feature of the Italian economy is the dichotomy between the Northern and the Southern regions. This dichotomy is reflected in several aspects ranging from socio-economic composition and productive basis to employment and urban structures (Hilferink et al. 1994). A large set of indicators may be used to assess such differences. We will focus here on a few aspects of the North/South dichotomy which are directly related to the mobility patterns and needs, by focusing in particular on the role of the urban structure as a background framework for this case study.

The Italian urban pattern has an interesting structure which has in recent years received a great deal of attention in both policy and research, especially by economists, geographers and demographers. In general, we may identify five metropolitan areas (Turin, Milan, Genova, Rome, Naples), three polycentric systems (Veneto, Emilio-Romania, Tuscany) and several minor cities. This already suggests the drastic contrast of urban density and other features in the Northern part compared to the Southern, as well as a prominent role of interurban relationships in Italy.

What emerges from several studies is also the prominent qualitative difference between the spatial pattern of interurban relations in the Northern urban cities (Celant et al. 1992). In particular, the Milan metropolitan area and the three Northern polycentric systems are actually evolving into a spatial ramification which has sometimes been described by 'the network paradigm' (Dematteis 1985). In other words, the spatial division of work and functions is less and less driven by the principles of hierarchical organization that inspired Christaller's central place model (Christaller 1933). Functions and activities located in the nodes of the urban networks are less dependent on demand thresholds and surrounding market areas, but rather on complementary relations with other nodes in a larger region. This tendency, which is actually driving the spatial evolution of many advanced countries, is rather dominant for the whole Northern Italy, while it is almost absent in the Southern part (Muscarà 1978 and RECLUS 1989).

In this context, a further analysis of the Italian spatial structure can be carried out by investigating some basic socio-economic indicators describing the nodes of the inter-urban network we are interested in (Table 4.9).

First of all, we will consider the indicator 'income per capita'. Table 4.9 gives a clear picture of the sharp difference between North and South. Average income per capita in urban nodes of the North is 19,047 millions liras, while it is only 13,053 in the South (1983). The role of income per capita is rather relevant for mobility. It is generally accepted that a clear direct relationship exists between income level and mobility (Blaas et al. 1992). Income per capita does not only give a measure of what may be called ability to pay, but also, given some plausible hypotheses on the evolution of consumption patterns, a clear indication about willingness to pay for personal mobility. The urban and spatial-economic structures in Italy also suggest something like a 'rich' satellite system around Milan, with at least three 'rich' corridors (Milan-Bologna-Ancona; Bologna-Padua-Venice-Trieste; Milan-Verona-Vicenza-Padua) and a rich cloud in the Tuscany system. Central and Southern Italy is from a socio-economic angle more homogeneous, showing lower income per capita levels. Around Naples we observe a sort of 'poor' satellite system.

Table 4.9. Demographic data of Italian cities with more than 150,000 inhabitants (average figures concern all cities of more than 20,000 inhabitants).

name	population (1991)	employed population (1991)	% employed in industry (1991)	% employed in trade (1991)	% employed in services (1991)	income per capita (x 1000 liras) (1987)	cars per capita (1990)
NORTHERN REGION							
TURIN	961,916	441,686	34.1	21.3	44.6	18.5	0.56
MILAN	1,371,008	765,808	25.4	24.2	50.4	21.3	0.59
BRESCIA	200,722	96,798	28.3	22.8	48.9	22.3	0.59
VERONA	252,689	100,690	29.3	23.9	46.8	18.7	0.56
VENICE	308,717	139,305	25.3	23.1	51.6	15.2	0.38
PADOVA	215,025	99,798	20.3	26.9	52.8	20.2	0.56
TRIESTE	229,216	82,383	16.6	26.0	57.4	17.0	0.53
GENOVA	675,639	234,103	23.6	23.1	53.3	15.8	0.48
PARMA	168,905	85,349	38.3	20.8	40.9	19.4	0.59
MODENA	176,148	86,212	36.2	22.6	41.2	19.4	0.63
BOLOGNA	404,322	207,097	21.7	21.8	56.5	20.2	0.59
Average	*168,629*		*26.9*	*23.1*	*50.0*	*19.0*	*0.55*
CENTRAL REGION							
FLORENCE	402,316	203,534	22.6	25.3	52.1	19.0	0.67
LIVORNO	167,445	57,251	19.1	24.6	56.3	13.6	0.56
ROME	2,693,383	950,309	16.2	21.8	62.0	15.9	0.59
NAPLES	1,054,601	300,573	16.3	20.1	63.6	10.8	0.40
SALERNO	153,436	46,423	18.7	18.4	62.9	16.2	0.48
FOGGIA	155,042	43,796	17.6	22.2	60.2	11.6	0.40
BARI	341,273	110,412	21.2	24.7	54.1	15.2	0.45
Average	*190,129*		*21.2*	*22.1*	*56.7*	*15.4*	*0.59*
SOUTHERN REGION							
TARANTO	232,200	73,080	37.1	17.5	45.4	11.9	0.42
REGGIO CALABRIA	169,709	36,316	8.8	25.9	65.3	12.5	0.45
PALERMO	697,162	161,966	14.3	28.9	56.8	10.9	0.48
MESSINA	272,461	63,825	13.9	23.0	63.1	11.9	0.45
CATANIA	330,037	97,524	19.7	25.4	54.9	11.9	0.53
CAGLIARI	203,254	85,914	16.4	21.6	62.0	14.4	0.56
Average	*159,354*		*17.9*	*22.6*	*59.5*	*13.0*	*0.46*

Source: Italian statistics (Capoluoghi di provincia).

Mobility patterns are also affected by demographic and employment compositions. Densely populated areas often have congestion problems, while spatial employment compositions affect daily mobility patterns, mainly commuting. With regard to the demographic dimension, the prominent position of central urban areas in Italy (Milan, Genova, Rome and Naples) clearly emerges. Once more the relevant Northern inter-urban patterns show also a high population density, while the Southern metropolitan system is less densely populated. Looking at employment compositions, the Northern nodes present on average a higher degree of industrial and trade employment, while the Southern parts are more unbalanced towards the service sector, basically public administration.

Car ownership may be regarded as a proxy for potential interest in telematics, since the number of cars per inhabitant is a measure of the ability to travel by car. By Table 4.9, the relationship between car ownership and the level of income can be investigated. Comparing the Northern region with the Southern region, it seems that the socio-economic contrast is also reflected by car ownership levels in these two regions. Surprisingly, such a relationship is not found when we look within the North. Car ownership in the Tuscany polycentric system is clearly higher than in Lombardia or Veneto, although these latter two regions have a higher average level of income. This is probably due to higher traffic congestion problems and to more reliable alternative transport modes in these areas.

4.3.2 Characteristics of Network Segments

The inter-urban road network considered in our case study consists of the Italian motorway network (Autostrade) and the network of main secondary roads (Strade Statali). The Italian motorways are mainly toll roads, with the exception of the Southern links from Naples to Sicily. Since drivers usually have more than one travel option, it is important to also consider the number of other alternatives than toll roads in order to evaluate the potential role of telematics in route choice situations. The spatial differences in road infrastructure over Italy are quite evident. In the Northern part of Italy the main and secondary networks are both extremely dense and widespread, so that in principle the one might be a reasonable substitute for the other. On the contrary, in the central and Southern part the two networks are more complementary, with the main roads connecting the bigger cities, and the secondary ones providing connections to the smaller centres.

In the following we will show some spatial variations in traffic conditions on the Italian motorway network, which are relevant for the use of telematics applications. We will focus on the following indicators: level of service (which is a proxy for congestion levels), traffic volumes, intensities of freight movements and accident rates. Unfortunately, reliable recent information on service levels for the whole network is not available. Therefore, levels of service levels from 1985 are used (see the map in Figure 4.10) which are based on United Nations traffic counts; in the case of the other three indicators more recent data are used from 1992 (Figures 4.11-4.13), which are based on data from the Italian AISCAT road authority.

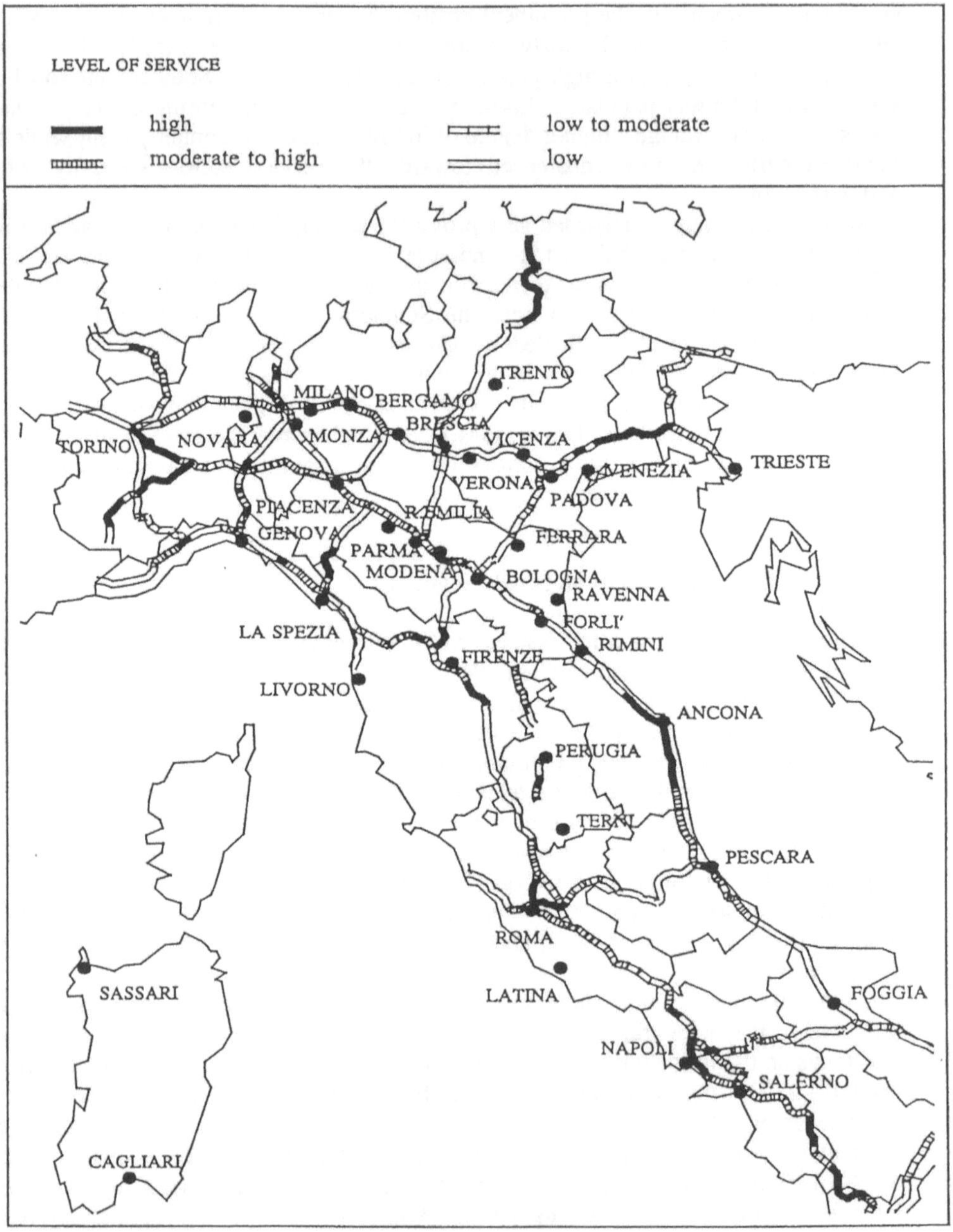

Fig. 4.10. Congestion levels on Italian motorways (1985).

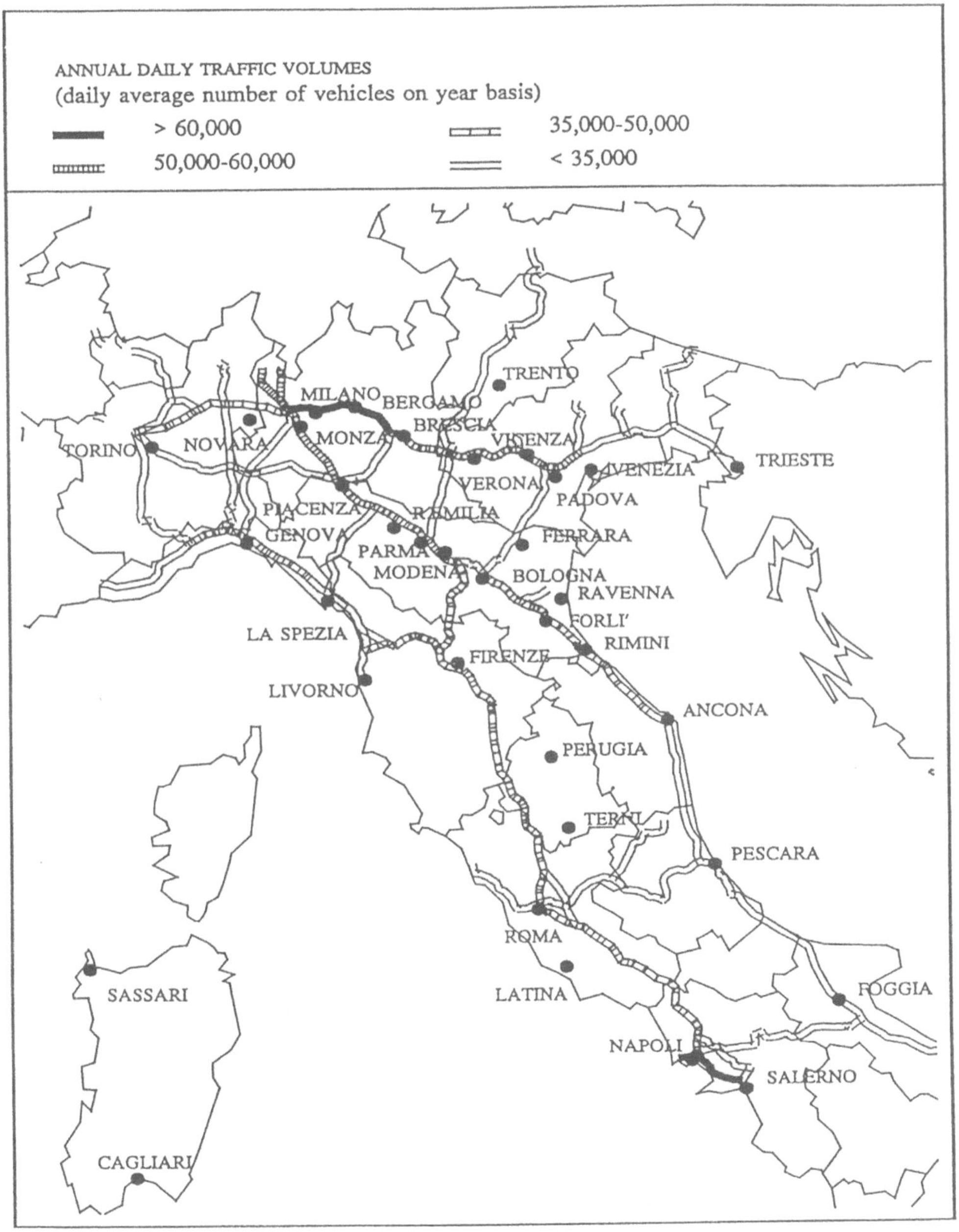

Fig. 4.11. Annual daily traffic volumes on Italian motorways (1992).
Source: AISCAT (1992).

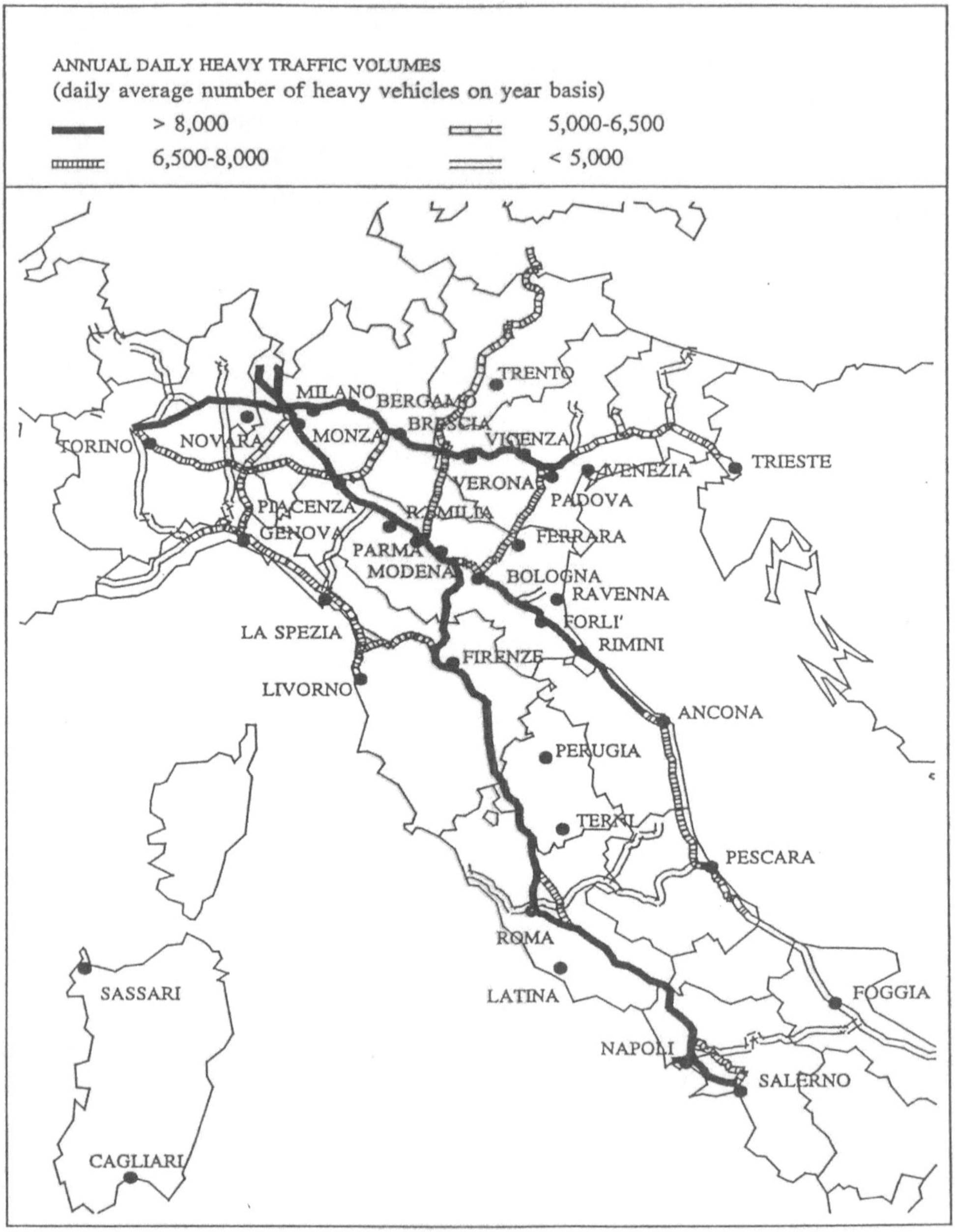

Fig. 4.12. Annual daily volumes of heavy vehicles on Italian motorways (1992).
Source: AISCAT (1992).

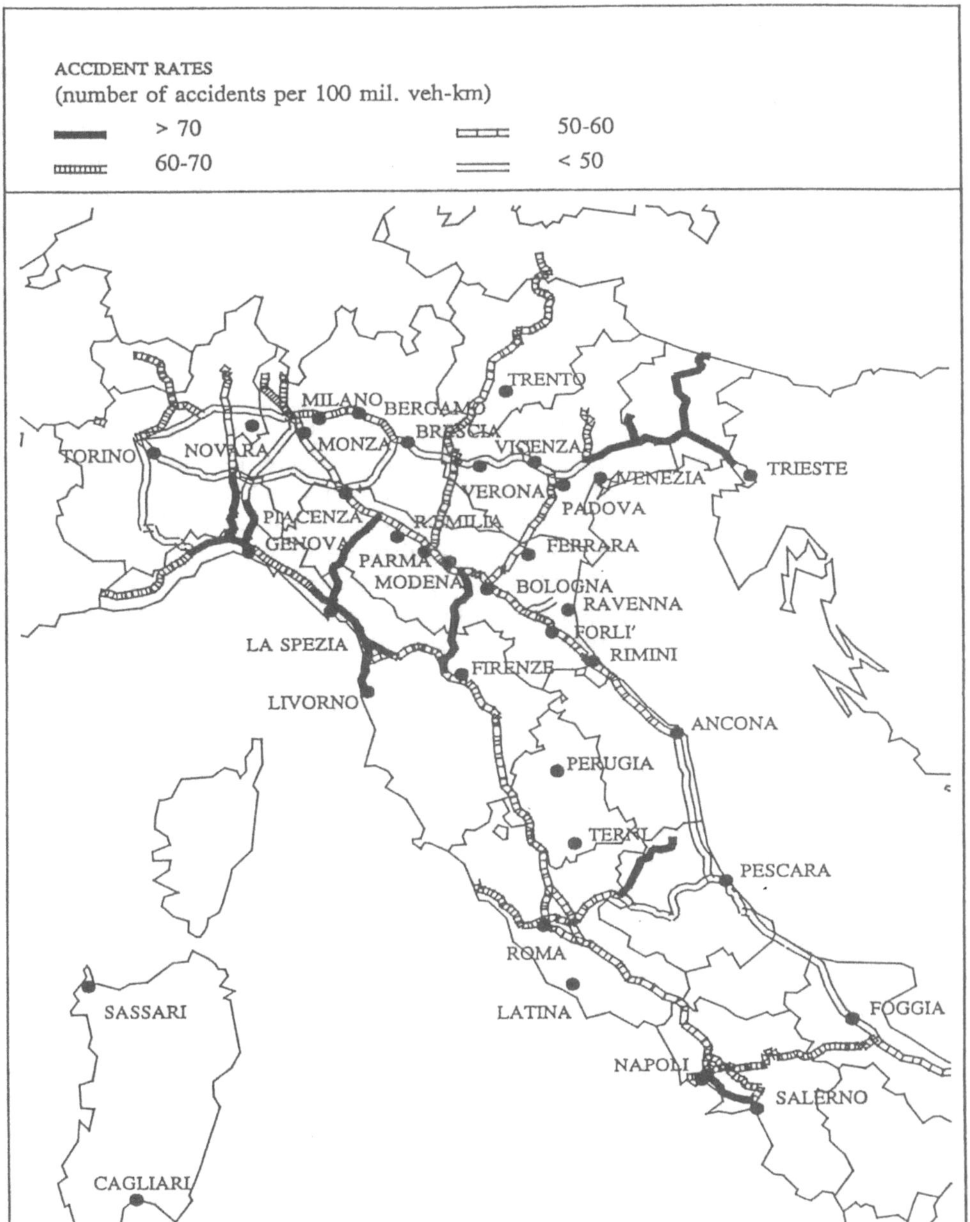

Fig. 4.13 Accident rates on Italian motorways (1992).
Source: AISCAT (1992).

Level of service

The level of service, basically a ratio of road capacity and traffic volume, may be interpreted as a measure for congestion problems on motorways (see MARTA 1993 and Small 1992) and related traffic safety (for an attempt to model the relationship between highway traffic density and safety, see Shefer and Rietveld 1994). Therefore, it reveals a potential need for telematics. Looking at Figure 4.10, we may point out the spatial variation of such problems, which seem to be concentrated on specific segments of the network. In the Central and Southern part of Italy the most heavily congested motorway segments are concentrated in a few sites: the ring around Florence, the Rome feeder system, the Naples feeder system and the Adriatic corridor around Ancona. In the Northern part, problems due to the level of service seem to be more evenly distributed. In this part of Italy two broad classes of cases can be observed: the Transalpine links and their feeders and some specific inter-urban segments.

Traffic flows

Daily traffic flows reveal the relative importance in terms of users' needs and also provide a proxy for the potential number of people that may be affected by the implementation of telematics devices. From the map in Figure 4.11 we can observe the importance in terms of traffic flows of the corridors and links feeding the Milan metropolitan area, especially the link Milan-Brescia and the link Naples-Salerno. Some other links play also a vital role in the Italian interurban network. These are in particular the integrated corridor Milan-Bologna-Florence-Rome-Naples (Autostrada del Sole), with a particular emphasis on the Milan-Bologna link, the corridor Milan-Venice and the corridor Bologna-Ancona. Summarizing, the following patterns appear to emerge:

* a high polarization of traffic gravitating around the Milan metropolitan area, a high intensity of traffic around Naples and a relatively minor relevance of the traffic situation in Rome, Genova and Turin;
* high traffic intensities on the Northern inter-urban corridors;
* an important role played by the Autostrada del Sole in connecting the Northern and Southern parts of Italy.

Flows of heavy vehicles

Traffic problems due to the presence of heavy vehicles are quite common for the whole Italian network. From the map in Figure 4.12 it can be observed that these problems are quite evenly distributed over the entire network. In particular, it can be noticed that almost every link of the dense Northern part of the network suffers from large flows of heavy vehicles, in combination with the Autostrada del Sole as well as the Bologna-Ancona corridors which operate as North-South traffic arteries.

Accident rates

Safety is one of the most pressing problems in transportation. There are therefore large expectations on the role that telematics can play in this context. Figure 4.13 provides a clear view on how (relative) numbers of traffic accidents are spatially distributed over the Italian network. From these maps it appears that safety is more related to other aspects (probably the physical lay-out of the road, local weather conditions and local driving habits) than to traffic volumes or levels of service. In particular, corridors with relative high accident rates appear to be Bologna-Florence, Parma-La Spezia-Livorno, Treviso-Trieste, the Genova feeder system and the corridor Teramo-Rome. Among these, only the Bologna-Florence corridor is both quite congested and unsafe.

From the above analysis of the characteristics of the Italian motorway network it is evident that the potential use of telematics systems may largely vary between various road corridors, depending on their geographical location and traffic conditions. However, as stated in our introduction, a most critical issue for the potential of telematics applications is the individual response and acceptance. Spatial variations in the telematics potential and use are dependent on the different ways individuals respond to informational stimuli in different situational conditions. In Chapter 6 of this book some hypotheses about the relationship between on the one hand drivers' needs for (dynamic) information and on the other hand conditions of the traffic environment are tested.

4.4 Conclusions

In this chapter we have in the first place tried to develop a systematic methodology for deriving a practical typological approach to the assessment of the potential telematics market from the viewpoint of infrastructural and local conditions and tested it for a sample of European road corridors. A full listing of European road segments from the viewpoint of problem cases and telematics potentials may help in building a more appropriate prioritization scheme for transport telematics investments in Europe, taking into account features that play an explanatory role regarding the relevance, use and adoption rates of transport telematics applications. In the second place, we have tried to describe and analyze regions and road corridors on a more local level in Italy. This case-study area offered in this respect quite some relevant spatial socio-economic contrasts. For example, looking at both traffic conditions and consumer market size indicators, it can be concluded that there is a significant difference in the opportunity for commercializing transport telematics devices between the Northern part of Italy and the Southern part.

In conclusion, there is much scope for transport telematics applications on the European road network, but this potential may vary between different corridors and between different regions. But the specific type of appropriate technology used depends on the morphology of the network, the local/regional conditions, the geographical coverage of transport corridors, the administrative support mechanisms, socio-economic conditions and user needs. This will be further elaborated in Part B which presents various case studies.

PART B

ACTOR CASE STUDIES ON TRANSPORT TELEMATICS

5 Telematics Information and Car Drivers' Behaviour: A Dutch Case Study

5.1 Introduction

Transport telematics serves to increase the efficiency in road use by offering information on road conditions, so as to induce behavioural changes among road users that would lead to a better capacity use. This chapter deals in greater depth with the impact of telematics on the transport behaviour of motorway users. The issue of information provision to individual road users and the aggregate results for the whole network has been studied quite intensively in the past years (e.g. for reviews see Bonsall 1992 and Emmerink et al. 1994a). We will first offer a concise overview of some findings by various authors (Section 5.2). Next, Section 5.3 of this chapter will present a case study in this field regarding VMS and RDS use in the Netherlands. In a separate Annex, we will also present empirical results from a modelling experiment.

5.2 The Impact of Telematics on Road Use: Review of Factors of Influence

In the past two decades several empirical studies on the factors affecting the drivers' choice of routes have been carried out. In this section the main findings of this research activity in various parts of the world will be summarized. Part of this inventory of findings is taken from Bonsall (1992) and Kokkota (1993).

In the inter-urban context, it is generally agreed that for most types of journeys drivers attempt to minimize travel time or distance. Between 75% and 90% of all choices are based on these criteria (Outram and Thompson 1978). However, a significant proportion of inter-urban travellers simply follow main roads or signposted routes, presumably because they are unfamiliar with the road network through which they pass (Wootton et al. 1981) or because they do not want to risk the uncertainty incorporated in transitional behaviour.

In the urban context the situation is less clear. In some cases it is concluded that time minimization is the dominant criterion, while in other cases importance is attached to aspects such as road type or hierarchy, avoidance of congestion, and avoidance of stops and traffic signals (e.g. Ben Akiva et al. 1984 and Ueberschaer 1971). In the European DRIVE I programme, route choice criteria of drivers were compared between four major cities in Europe (CARGOES, 1990a,b and c). Route choice criteria appeared to vary between the cities and according to different journey purposes even when the possible influences of cultural differences and different network conditions were discarded. The questionnaire on which these results were based distinguished between a number of time related criteria and it is interesting to note that, although simple time minimization was important for all types of journeys, in no case was it the main criterion for most respondents; criteria like avoiding congestion and certainty of arrival time appeared to prevail.

Several attempts have been made to assess whether drivers succeed in selecting routes which achieve their stated objectives. This seems to be very difficult, because most criteria are subjective or time variant and thus difficult to measure objectively and accurately. Nevertheless, from a study of commuters' route choice in the Santa Monica Freeway Corridor of Los Angeles it appeared that in 'normal' conditions most drivers were successfully minimizing time, but were not adjusting successfully to situations of non-recurrent congestion (Al-Deek et al. 1989; see also Verhoef et al. 1994). Such a finding highlights the potential value for drivers to have access to real-time traffic information.

Research has also been carried out on drivers' perceptions and the use of alternative routes (focused on commuters). On this subject ambiguous results were found. In some cases it was found that, although there were route alternatives available, drivers tended to use day by day the same route due to little knowledge of the alternatives (e.g. Wright 1976). Other studies however showed that considerable shares of commuters used routes they were not familiar with on one or more occasions and selected these on-route in the light of prevailing road conditions (Bonsall and May, 1986). In general, the choice of routes seems to be dependent on recent experiences (Yang et al. 1993), and the availability of traffic information plays an important role in breaking the strong influence of habitual patterns on individual route choice (Van Berkum and Van der Mede 1993).

Conceptual models of drivers' route choice

Results of empirical studies cited above have led to the formulation of conceptual models of the process of route choice. Important features of such conceptual models include the distinction between strategic planning (based on expected network conditions) and tactical adjustment (in the light of conditions actually met), the existence of temporal disequilibrium (due to inertia in drivers' adjustment to new conditions), bounded rationality, attitudes to uncertainty and complex search strategies. More recently developed route choice models are found in the context of assignment modelling (see Ran et al. 1994). They are mainly concerned with predicting realistic flows on links rather than route choices. An important feature of such models is the representation of the interaction between link travel times and flows: fast links attract traffic and consequently become slower in the modelled process. Most of these models are mainly concerned with predicting average conditions over a period of time rather than actual conditions on a particular day. Many of them seek to generate equilibrium flow patterns which might be expected to result after a period of time. A widely used test of the success of such models is the achievement of a Wardrop equilibrium solution, wherein no driver can unilaterally reduce his travel cost by modifying his current routing pattern. The most sophisticated modelling approaches in assignment modelling also incorporate network dynamics (wherein flow and congestion change over time), stochastic choice (which recognizes differences in the perception or levels of knowledge of link costs between different user groups), day-on-day variability and a distinction between equipped and non-equipped drivers (e.g. Emmerink et al. 1994a).

There is a great need for evidence of real-world empirical evidence on drivers' behaviour to transport telematics information. Not so many efforts to study revealed behaviour have been carried out in the past (amongst others Turner et al. 1978; Dudek et al. 1982; Owens 1980; Khattak et al. 1991; Caplice and Mahmassani 1992; LISB 1991). In the next section, we will present results of a recent empirical study in the Netherlands to investigate this real-world drivers' behaviour.

5.3 A Case Study in the Northern Wing of the Dutch Randstad

5.3.1 Context of Site and Description of VMS Application

Like most of the European capital cities, the area around Amsterdam suffers from severe traffic jams on its surrounding major roads. The major roads connecting Amsterdam with surrounding towns are heavily congested, especially during the peak-hours. Two main developments are to blame for this. First, companies moved out of the inner cities to the suburbs, where large office buildings were constructed in the past decade. The South-Eastern, Southern and South-Western parts of the agglomeration were favourite locations for these companies. Second, government policies to guide the suburbanization went in another direction. Residential locations were planned at some distance from the large cities in the form of growth towns. Especially at the Northern side of Amsterdam, large numbers of dwellings were constructed. However, the growth of employment in these new cities stayed far behind the expectations of the government. This led to an increasing imbalance on the local labour market and consequently to an increase in commuting distances (Bruinsma 1995).

As a consequence, one of the main traffic problems is the crossing of the river IJ which splits Amsterdam just north of the inner city into two parts. Every day, a large flow of commuters travel from the residential areas north of the river IJ to the employment centres in the southern part of the agglomeration. But other access roads to Amsterdam also have severe daily traffic problems. At various locations during non-peak hours severe congestion occurs, for instance, as a result of minor incidents.

These developments caused the need for the completion of the orbital motorway in the Amsterdam region, which is a very important link for all regional and through motorway traffic (Buijn et al. 1994). Major parts of the western and southern side were already completed in the 1970's and 1980's. In September 1990 the last part of the Amsterdam orbital motorway - the Zeeburgertunnel under the river IJ - was completed, opening up the Northern and Eastern segments. Its main functions are to improve the accessibility of Amsterdam and the province of North Holland to the region north of the river IJ, and relieving the secondary road network in the Amsterdam region.

The completion of the orbital motorway provided new routing alternatives for a considerable number of users of the regional inter-urban road network, as it opened up new opportunities and capacity for road users along the eastern and northern side of Amsterdam. A dynamic traffic management application consisting of Variable Message Signs has been implemented to support users of the ringroad in selecting their route. The system is called Route Information Amsterdam (RIA). It is expected to help optimize traffic flows on the ringroad by making a balanced use of the ringroad. RIA provides users approaching the ringroad with information on traffic queues (including the length of the queues) and with information about closure of tunnels or driving lanes. The information is generated automatically from data from the Motorway Control and Signalling System (MCSS) in Amsterdam, and is presented on the sign in fixed display formats. If there are two queues close together, the lengths of both are added and displayed as one queue. If there are two or more different queues, only the first queue to be encountered is displayed. The value of the information is lower when there is a longer distance to go, so that the first queue to be met should be the one indicated.

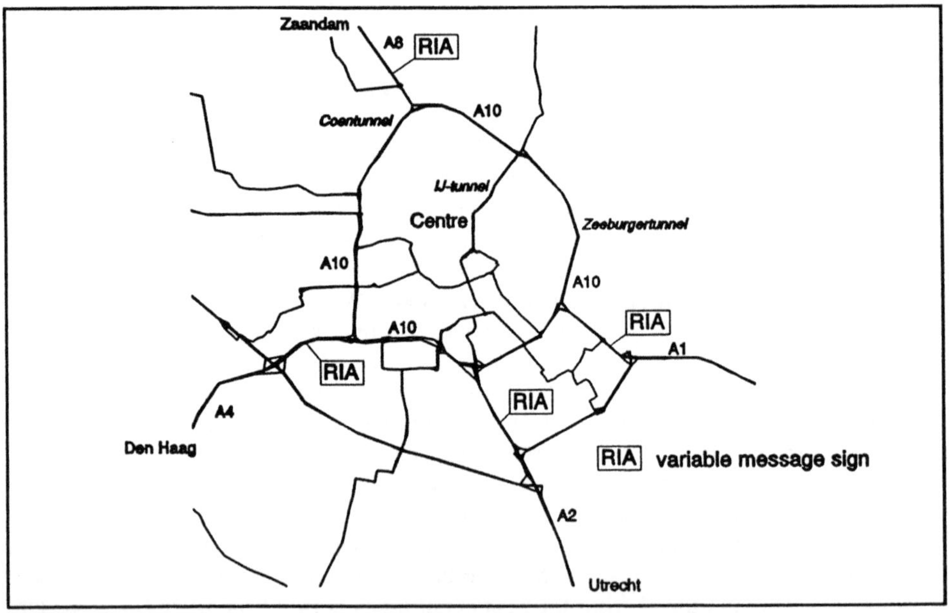

Fig. 5.1. Road network and VMS signs in the Northern Wing of the Randstad.

The type of information provided is specifically meant for those road users who are familiar with the network, knowing their route possibilities when passing the VMS signs. Furthermore, the longer the distance driven on the ringroad, the less discriminating the alternatives to make a choice for a clockwise or anticlockwise movement over the ringroad. It is thus expected by the road manager that the information is useful for through traffic and a certain part of traffic with Amsterdam as its destination.

Some technical features of the VMS signs are: the signs are mounted in a box of 8 m long and 2 m high; each of the two lines of text consists of 21 characters; and the character height is 0.35 m, large enough to allow sufficient time for road users to read the signs. On a sign two lines of text are displayed (in Dutch). The upper line concerns the traffic situation in one direction of the ringroad from the respective VMS site, the bottom line in the other direction, so that frequent users always read 'their' line at the same position. There is a consistent sequence of the items in a line. The first word is the direction concerned, the next item the length of a queue (if there is one). Then the information is completed by 'km queue'. Also an arrow is added to indicate the motorway driving lane(s) to be taken for the respective ringroad direction. If no queues occur, the message 'no queues' is displayed. Except for these standard messages, other kinds of messages (like closed driving lanes) can also be given, since the control centre is fully flexible in displaying texts.

In November 1991 the first variable message sign was put into use at the most strategic location, namely on the motorway A8 from the North before the junction with the ringroad A10. In April 1994 another three identical signs were installed respectively on the access motorways A1, A2 and A4 from the South, each just before the respective junction with the orbital motorway (see the map of Figure 5.1). At present the system consists of four VMS signs in full operation.

The Dutch Ministry of transport has decided to extend the system in the future, along with other motorway traffic applications. The traffic data collecting system will be implemented on other motorways around Amsterdam, by which an extension of the RIA VMS signs will be possible. Another 15 signs will be installed south of the city and on the ringroad itself (the latter for a better tuning with local traffic from the city).

5.3.2 Survey Methodology, Design and Set-up

A behavioural survey was carried out among users of the motorway network approximately three months after the full implementation of all four variable message signs. The survey target consisted of car drivers visiting Amsterdam via one of the four main motorway access roads where VMS signs have been installed, and who made use of the ringroad. The survey had a sample size of 826 observations. The revealed character of the survey made it possible to investigate in-depth the market potential and the response elements from the previously presented nested approach via a segmentation of the survey sample and the measurement of attitudes and actual behavioural changes regarding this operational telematics system (Chapter 3).

It was hypothesized that important explanatory segmentation variables (identified in the Nested Approach) should be sought in the age, gender and income/social group of the respondents as well as their travel characteristics like their experience with dynamic driver information, frequency of travelling (and inherently the familiarity with alternative routes) and trip purposes. Looking at the possible effects of the kind of information provided on travel behaviour in the context of this site, the emphasis was given on choice options related to the route followed and the departure times of the trips. The relatively long-distance character of the car trips made implied that it was unlikely that there would be an impact on other travel behaviour parameters (like changes in destination choices or trip rescheduling).

Several considerations underly the possible impact of the information on route choices. For instance, the change from a planned route under the influence of dynamic information may depend in the first place on the existence of any possible alternative routes available in the specific situation. Secondly, the demand for alternative routes may be determined by the (expected) duration and cause of the queueing ahead on the motorway being used. Personal preferences of motorway drivers with respect to traffic delays and rerouting and travel features like distance and time restrictions play an important role here. If, for example, the expected delay ahead on the followed route is equal to the detour time of any other possible route, an as yet unknown share of drivers may change route. This type of reaction however is more likely to take place when the travel distance is longer and the same detour distance is relatively shorter. The uncertainty which stems from using such an alternative route may also prevent an unknown share of drivers from taking this type of action, particularly those who are less familiar with the area (BATT 1994).

In order to obtain statistically valid results, a representative sample of respondents for the survey was necessary regarding the time of day of driving at the VMS locations concerned. Such a random selection was made by selecting drivers stopping at petrol stations on the respective four motorways in the neighbourhood of Amsterdam. It is recognized that by this strategy the survey population may be slightly biased, for example, to those not incurring petrol costs in the area (like, for example, business drivers), given the slightly higher petrol price at stations along motorways. Four gas stations were selected for the distribution of questionnaires. These questionnaires could be returned by post, from those drivers fulfilling the criterion to less or more frequently driving to or around Amsterdam (known by asking them). This distribution took place within one week, during whole days between 07.00 in the morning and 20.00 in the evening. Thus, in this period both drivers in peak-hours and non-peak hours were selected.

5.3.3 Driver Profile

Respondents' socio-economic profile

About 44% of the respondents are at the age of 25 to 34 years (see Figure 5.2). The age categories 35-44 and 45-59 make up 27% and 23% of the total, respectively. The young ones (24 years and younger) and the elderly (60 years and older) are a minority in the survey - each group forms about 3% of all respondents.

There were significantly more males than females in the sample; 84% were male. It was remarkable that the female respondents were in general younger than the males. From the female group 66% were younger than 35, while for the male group this percentage accounted for 44%.

About 8% of the respondents had their driving licence for a period of up to 5 years, and 21% for 6-10 years. Approximately 71% of the respondents have been in the possession of a driving licence for more than 10 years; 36% for 11-20 years, 24% for 21-30 years and 11% for more than 30 years. It may be concluded that the majority of the respondents had a long driving experience.

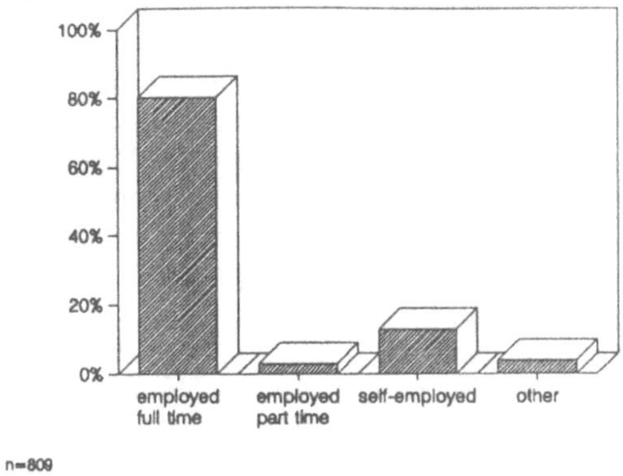

Fig. 5.2. Age categories of respondents (% of total sample).

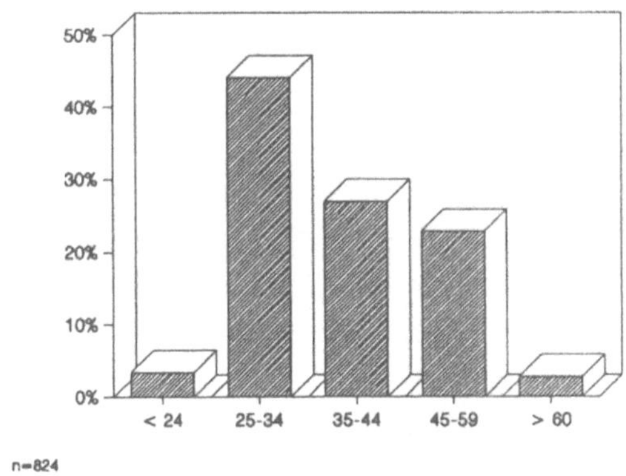

Fig. 5.3. Working status of respondents (% of total sample).

The majority of the respondents, 80%, had a full time job (Figure 5.3). Only 3% were part time employed and 13% were self-employed. Other working status categories (housewife/househusband, unemployed/seeking work, student, retired and other) accounted for the remaining part (4%). In short, the population surveyed may be characterized as mainly male full time workers whose age is between 25 and 60, and who had a long driving experience.

Travel characteristics

The following features related to the car trips were investigated: the trip purpose, frequency of driving, time of day, flexibilities in arrival time, travel distance and -time, number of persons in the car and the availability of route alternatives.

An important feature was the usual purpose of the trip. Business appointments were mentioned by nearly 36% of the respondents as one of the dominant purposes of their trips by car in the direction of or along Amsterdam (Figure 5.4).

Approximately 55% have as their usual journey purpose travelling to or from their work (commuters) or travelling as part of their work (services and freight/delivery drivers). Drivers who make only discretionary trips (leisure, visiting friends/relatives, shopping/personal business and other purposes) make up the remaining 9%.

Business drivers were in general younger than discretionary drivers: 73% of the former group were younger than 45, while this share was only 55% for the latter group. Furthermore, there were relatively more females than males making discretionary trips (of the female group 21% and of the male group 7%), while business drivers had a relatively stronger representation in the male group (40% of this group) compared with females (16% of this group).

When asked how often the respondents make use of the ringroad, nearly half of them (48%) said five days a week or more (see Figure 5.5). About 17% make use of it three or four days a week, 18% one or two days a week, and 14% less than once a week, but more than once a month. Only 3% of the respondents drive over the ringroad less than once a month. It may be concluded that most respondents were frequent ringroad users.

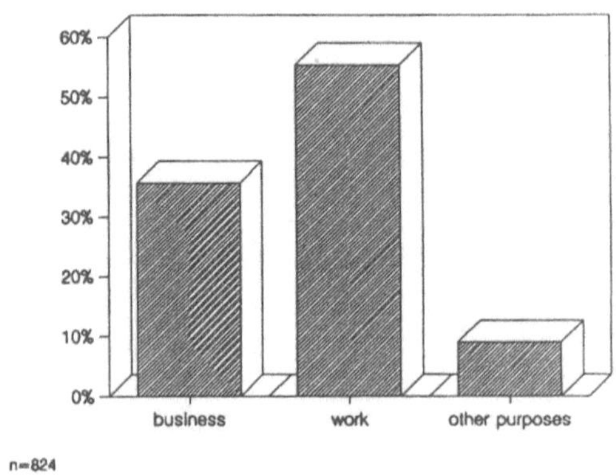

Fig. 5.4. Usual trip purpose (% of total sample).

As could be expected, from the motorists driving daily to Amsterdam (five days a week), the largest part was made up of commuters (84%). Those who drove only rarely to Amsterdam are dominated by discretionary drivers (62%). Drivers who mainly travelled for business purposes are equally represented in all classes of driving frequencies.

More than half of the respondents (53%) made use of the ringroad only during peak hours (7.00-9.00 in the morning and/or 16.00-18.00 in the afternoon), and about a quarter of them only in the hours between these two periods. The other respondents (24%) used the ringroad both in normally congested and non-congested hours of the day.

The usual travel distance of the trips (one-way distance) was for just over half of the respondents (53%) more than 50 kilometres. About 35% usually made journeys of 25-50 km, 11% of 10-25 km and 1% make trips of less than 10 km.

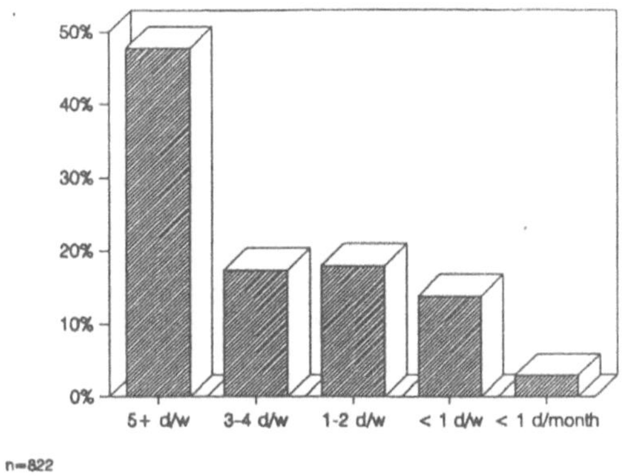

Fig. 5.5. Frequency of car trips made (% of total sample).

Most respondents had an average travel time (one-way travel time) between 20 minutes and 2 hours: approximately 32% travelled 20-40 minutes, 36% travelled 40 minutes to one hour, and 26% travelled 1 to 2 hours (Figure 5.6). Very short travel times (less than 20 minutes) and very long travel times (more than 2 hours) account for only 2% and 5% of all journeys, respectively.

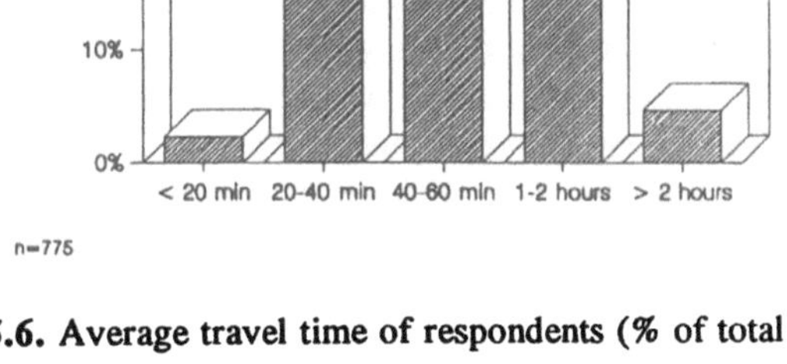

Fig. 5.6. Average travel time of respondents (% of total sample).

Nearly half of the respondents (47%) declared that they did not have any flexibility regarding their arrival time at the destination/work place. For 8% it was possible to arrive up to 10 minutes later than planned, and for 21% 11-20 minutes. 16% might arrive 21-30 minutes later, while about 8% were free to arrive more than 30 minutes later than the planned arrival time.

The great majority of the respondents (84%) usually drove alone when making use of the ringroad around Amsterdam. 16% travelled with one or more passengers.

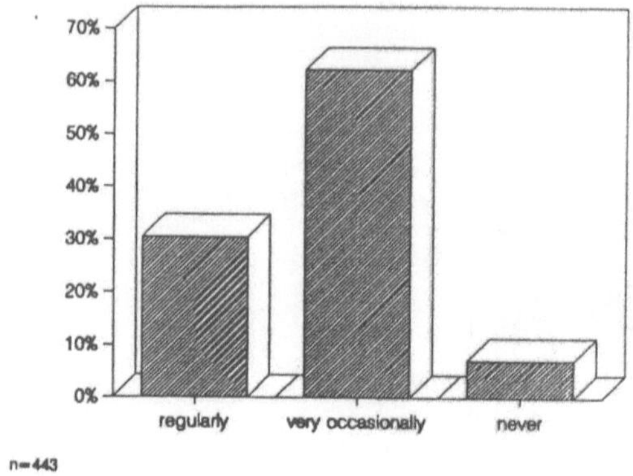

Fig. 5.7. Use of alternative routes (% of those who have real alternatives).

It appeared that 55% of the respondents were able to follow more than one route without a significant time delay to their destination. About 41% declared to have no real alternative(s) and another 4% did not know. Those drivers having route alternatives were also asked whether they ever made use of an alternative route. Of these respondents approximately 31% regularly took an alternative route, 62% only rarely and 7% never did (Figure 5.7). Thus, a significant number of drivers (16% of the whole sample) showed a non-steady behaviour with respect to the followed route for their trips. This high route choice flexibility and use of more than one route is in principle encouraging in terms of the potential usefulness of the driver information provided.

5.3.4 Attitudes to VMS

Almost all respondents understood the messages displayed on the electronic signs along the motorway. To 87% of them the information was clear, and for about 12% the information was quite easy to understand. For only 1% was the information difficult to understand (e.g. the information is unclear or there is not sufficient time to read it).

The great majority of the respondents was pleased or very pleased with the provided information: 35% and 55%, respectively (Figure 5.8). Only a low share (10%) found the information unsatisfactory. This latter group (93 persons) were also asked to give a reason for this. The main arguments given were:

- information was of no use because route cannot be changed (19%);
- information was not specific enough (unknown which exits can still be reached behind the tail of the queue) (15%);
- information was not always up-to-date (12%);
- the signs had been installed too close to the critical junction (11%);
- the information was difficult to understand (by poor readability etc.) (8%);
- information was not informative (6%);
- information did not tell which way is faster (5%).

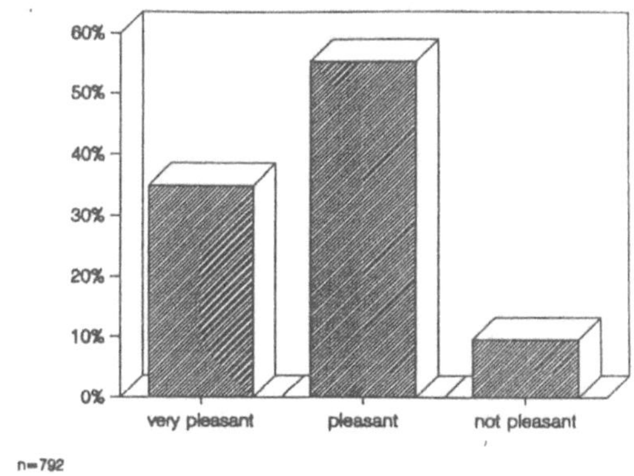

Fig. 5.8. Pleasure obtained from the dynamic queue information (% of total sample).

Approximately 21% of the respondents thought the queue information provided was always correct. The majority, 71%, thought the provided information was in most cases correct, while 8% had the opinion that the information was only sometimes correct. The number of respondents who claimed that the information was never correct was negligible. These figures indicate that the majority of drivers might not have a full confidence in the information, because they had previous experience of a wrong message.

5.3.5 Responses to the VMS Traffic Information

By far the most important question of the survey was whether respondents were ever affected in their choice of route by the provided information on traffic queues. The results are shown in Figure 5.9. In 28% of all cases the information had never affected the route choice of the drivers. The remaining 72%, however, claimed to have ever changed their route: about 40% only rarely, 23% regularly and 9% even often. It may be concluded that a really significant number of drivers had been affected by the information in their route choice since the VMS were installed.

Of those who often changed route, 37% found the information always correct and only 7% found the information only sometimes correct. For those who declared never to have been affected in their route choice, these respective figures were 23% and 12%. This variance may emphasize the negative influence of a reduced confidence in the system by experiences with wrong information.

Those respondents who have been affected by the information were asked some questions about the route they took as an alternative. On average, the length of this other route was the same as the planned route in about 8% of the cases.

Nearly 24% drove up to 2 km more, 43% drove 2 to 5 km more and 26% made a detour of more than 5 km (Figure 5.10). The alternative route was (usually) also by motorway for just over half of the route switchers (56%). The remaining alternative routes (44%) were (partly) off the motorway. The high number of route switchers using secondary city roads suggests that the information, of which the main goal is to optimize traffic between clockwise and anticlockwise directions on the ringroad, might also potentially generate some negative side-effects that might conflict with targets to keep motorway traffic as much as possible outside urban areas.

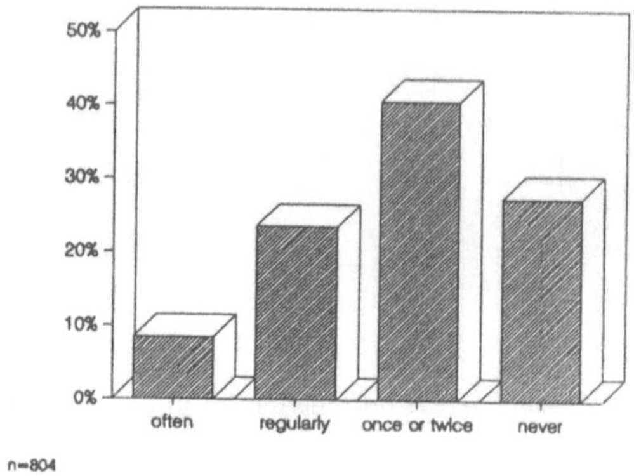

Fig. 5.9. Impact on driver's route choice of the VMS information (% of total sample).

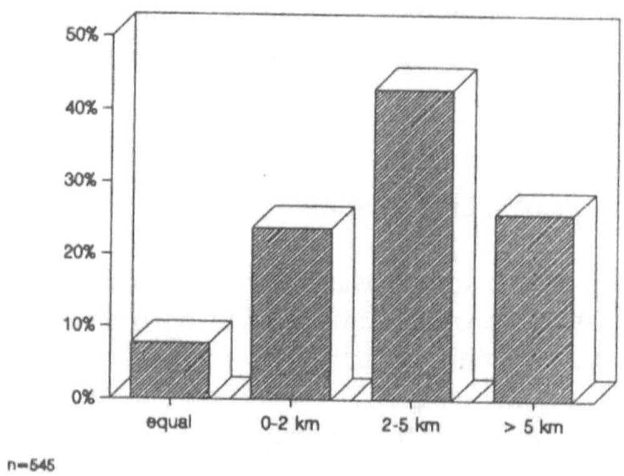

Fig. 5.10. Length of detour distance when changing route (% of route switchers).

The extent different VMS stimuli had impacts on route choice (Figure 5.11) was also investigated. Of those drivers being affected by the information, 29% declared to have taken this decision once or more when seeing a message of a queue length between 0 and 2 km. About 51% took this decision only when seeing a message of at least 2 km queue. For 16% a minimum queue length of more than 4 km was reason to change route, while only 4% changed route when queues of more than 6 km were indicated.

Approximately 34% of the respondents who ever changed their route due to the information provision declared to be able to evaluate how much travel time they might save in the most favourable cases. From this subgroup, about 31% could save up to 10 minutes, 50% of them could save up to 20 minutes, and 19% could save more than 20 minutes.

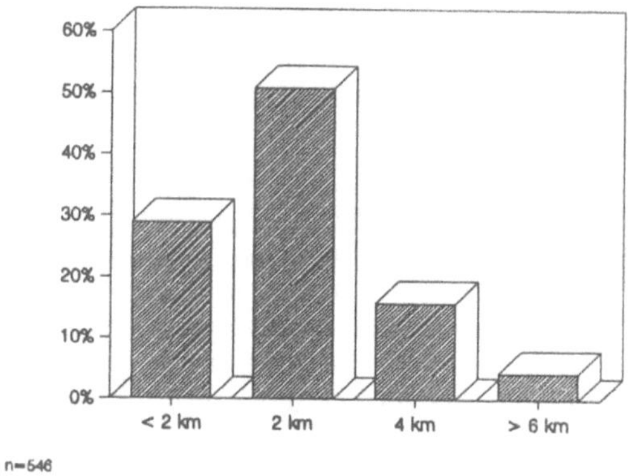

Fig. 5.11. Route change response to shown queue lengths (% of route switchers).

The satisfaction rates with alternative routes were encouraging. Nearly 38% of the route switchers declared to be better off with the other route, 13% felt that they were not better off. Almost half of the respondents (49%) did not know. Quite remarkable was the finding that among the drivers who were able to compare the detour route with the planned route, those who made detours off motorways were relatively less satisfied. From those staying on motorways felt 43% to be better off compared with 30% of those using a secondary road.

Besides the impact on *route choice*, travel behaviour is also affected regarding the *departure time* of the driver. It may be hypothesized that in some cases habitual behaviour related to departure time choice may change (for example, by a 'return to the peak' effect), because drivers may get used to a reduction of uncertainty by the information provided or, more directly, because individual travel times may be repeatedly reduced. Figures show that from the whole population surveyed, most respondents (82%) were never affected by the system in their chosen departure time. Respondents changing their departure time occasionally and regularly made up nearly 11% and 8%, respectively. When asked about the precise motives for these changes, it appeared that of those who changed their departure time, 79% declared to have done this for such reasons such as anticipation of traffic queues and avoidance of peak hours. This suggests that the majority of this group did not interpret the issue correctly. This implies that in general the on-route traffic information by the VMS itself had a neglible direct impact on departure time choices. Only 7% of the mentioned group (six respondents) based their departure time changes on traffic information, but in combination with supplementary information by radio.

The VMS system concerned is a public system implying no cost of the information provision to the end-user. In order to obtain an indication of the potential market for individual systems providing the same type of traffic information, the willingness to pay was investigated for in-car systems continuously providing this kind of information. It appeared that the willingness to pay for the information was in general quite low

(Figure 5.12). Almost half of the respondents (49%) were not prepared to pay anything. Maximum acceptable prices of 10, 20 and 40 ECU per annum accounted for 19%, 15% and 10% of the respondents, respectively (1 ECU is approximately 1.25 US dollar). Only about 7% were prepared to pay more than 40 ECU per annum.

It is interesting to note that these results on the willingness to pay are of the same order of magnitude of the results of driver surveys in other countries (see e.g. MARTA 1994a,b). In the UK the willingness to pay for, inter alia, the 'Traffic Master' system has been investigated. This system provides the same sort of information as RIA, also on a continuous basis. It appeared that also in the U.K. case about half of the respondents did not want to pay anything for this information (Banister 1995).

The ones who indicated that they found the information very pleasant showed a relatively high willingness to pay. About 47% of those willing to pay more than 40 ECU per year found the information very pleasant; for those who did not want to pay anything, this share was 27%. Looking at actual route change, it also appeared that those often changing route showed a higher willingness to pay.

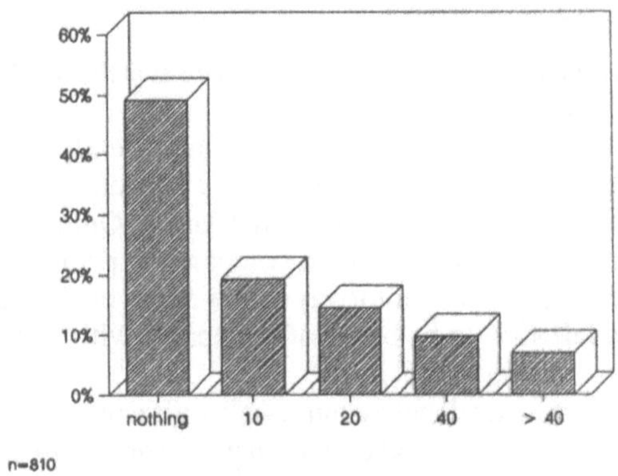

Fig. 5.12. Willingness to pay (prices in ECU per annum; % of total sample).

From those willing to pay more than 40 ECU per annum, 16% often change route while 14% never do. From those not willing to pay anything, only 6% often change route, whereas 36% never do so. The still considerably high share of drivers who never changed route in the group willing to pay more than 40 ECU per annum, implies that not only the actual usefulness of the VMS information in terms of saved travel time, but also other benefits may influence the willingness to pay for this type of information.

5.3.6 Impact of Radio/RDS Traffic Information and Comparison with VMS

A part of the questionnaire concerned the experience and response of the questioned drivers to dynamic traffic information by radio traffic broadcasting (which is the existing means for disseminating dynamic information on queues occurring on motorways throughout the whole country, mostly at fixed times) and/or by means of RDS.

It appeared that the majority of the drivers (76%) regularly or often listened to this kind of traffic information (38% often). Those who listened to the radio were asked how they perceived this source of traffic information. A group of 21% thought the information was very relevant. 64% found the information just pleasant, while 15% stated to have no opinion. Almost half of the listeners (45%) had a neutral opinion about the reliability of RDS, while another 12% found the information unreliable. This result clearly indicates that there would be a need for more accurate information, like the information provided by the VMS.

Changes in route choice instigated by radio/RDS traffic information were somewhat lower than in the case of the VMS information. The information did make 65% of the listeners change from the planned route only once or twice and another 22% never. The satisfaction with alternative routes followed were of the same order as in the case of the VMS.

The ones who were influenced by radio traffic information in their route choice were more sensitive to route change by the VMS than others. It appeared that of those who regularly or often changed route owing to the VMS information, 90% once or often also rerouted their trip owing to traffic information provided by radio, while for those never changing route by the VMS this percentage is 48%.

Furthermore, the willingness to pay for the continuous availability of the VMS information was relatively higher among those who were frequent listeners to radio traffic information. From those never using radio traffic information, 37% were prepared to pay a certain amount for the VMS information, while this percentage was higher (56%) for those often using radio traffic information.

These results indicate that the information provided by the VMS is mainly used by those drivers who have experience with and already use other kinds of traffic information. For this group of drivers, the VMS information might be a substantial complement to these other sources of information, since if the VMS information did not add any value, then drivers listening to radio traffic information ought not be willing to pay anything for the additional RIA traffic information. Consequently, the level of acceptance of a new invention may also be complementary to the use of a related invention.

5.3.7 Influence of Socio-economic Characteristics

In order to identify the main explanatory variables that are related to user groups and characteristics of car trips made, and which might influence various kinds of responses to the information provided by the VMS, a cross tabular analysis will be applied in this (and the following) section. Given the context and nature of the survey, which concerns a cross-section measurement after the installation of the VMS signs, and given the nature of the questions, cross tabular analysis seems to be the most efficient analytical tool to assess these possible relationships. In addition to this method, also an econometric approach will be applied in the Annex to this Chapter.

A series of cross tabulations was constructed that covered all possible relationships between the variables investigated. These included on the one hand the correlation between VMS responses and socio-economic characteristics, and on the other hand correlation between VMS responses and trip characteristics. Hereafter we will describe correlations that have a sufficient significance level (below 5% using the Pearson method). The conclusions will be summarized in the following two sections. Section 5.3.7 investigates to which extent user responses depend on segmentational variables and Section 5.3.8 investigates the influence of travel characteristics, following the Nested Approach.

An analysis of the influence of socio-economic variables like age, years of having one's driving licence, working status and gender on response rates to the VMS showed that neither of the first three showed significant correlations with any type of response. In contrast to these variables, significant correlations were found for the gender variable.

The share of females who claimed to change their route often by the VMS information was 5%, while in the case of males this share was higher (9%). In contrast, the share of those who claimed never to change route were in the case of females 35% and in the case of males less (26%). In addition, a slight difference between the two sexes was found for the type of message causing the route change. It seemed that males were more sensitive to relatively smaller delays. 81% of the males were influenced by messages with a queue length up to 4 km, while this percentage was lower for females (72%).

Explanations could probably be found in psychological differences between the two sexes, but also here some correlation occurred between the gender and the trip purpose, because it was found that men travelled relatively more for business purposes than women (Section 5.3.8). This difference between males and females confirmed the results of other behavioural studies (e.g. Mannering et al. 1993).

There seemed also to be a small difference between the two sexes in regard to the willingness to pay for the continuous availability of the information at any time in the car. The female respondents are less prepared to pay for having VMS information than the male respondents. About 61% of the women do not want to pay anything, against 46% of the men. The male respondents are not only more willing to pay, but they are also willing to pay more. From the men 12% are prepared to pay a maximum amount of 40 ECU per annum, while only 2% of the women are! For paying more than 40 ECU per annum, these figures are 8% and 2%, respectively. One of the reasons for this difference between men and women might be a greater interest of men in technical gadgets.

5.3.8 Influence of Travel Characteristics

It is hypothesized that certain features related to the type of car trip one is used to make, may have impacts on the way one values and responds to on-route dynamic traffic information. As discussed in Section 5.3.2, the availability of route alternatives for example is a prerequisite for any use of such information. In the following analysis, the extent to which such characteristics may indeed have any influence is investigated.

The focus is on the following attitudes and responses to the VMS information: the level of satisfaction obtained from the information, the perceived correctness of the information, route changes and detour distances, change of departure time and the willingness to pay for the information.

Satisfaction from the information

A slight correlation was found between the level of satisfaction obtained from the information and the availability of route alternatives. From those drivers having a choice of more than one route (without a significant difference in travel time), 39% stated to find the information very satisfactory. This percentage was slightly lower for those not having such a choice (30%). However, in both groups high percentages of respondents found the information either satisfactory or very satisfactory (similar rates of about 90% for both groups).

This would mean that, although those who have real alternatives benefit more, all drivers would in some way benefit from the provided information, even if they cannot actually adjust their trip to it. An explanation can be found in a reduced uncertainty with regard to the expected traffic situation, which increases the 'well-being' of the car driver.

Perceived correctness of the information

No significant correlations were found in the case of the perceived correctness of the information. This conclusion seems plausible in considering the quality of the information on the VMS signs.

Route changes and detour distances

When looking at the impact of the VMS on the route choice of drivers, it seemed that some trip features indeed had some influences.

First of all, as was to be expected, the impact of the VMS information on route choices of drivers is clearly dependent on the quality of the set of alternatives available to them. From those in the sample who stated that they could follow more than one route with comparable travel times, 41% do often/regularly change route, while, for those stating that they did not have close alternatives, this share was only 22%.

In order to detect possible variances between drivers with different trip purposes, three groups were distinguished: (1) drivers who indicated to travel regularly for business appointments, (2) commuters whose main purposes were to travel to the work place and work-work (e.g. delivery) drivers and (3) drivers who indicated only to travel for discretionary purposes like leisure and shopping. The assumption behind this classification is that these respective groups might have a decreasing value of travel time, which might have an impact on the response to dynamic queue information.

Indeed, a significant correlation was found between the frequency of route changes and these trip purposes. The respective shares of business drivers, commuters and discretionary drivers who often changed route by the information provided were 10%, 8% and 6%. In the case of never changing route, these shares were 20%, 31% and 36%. Thus business drivers seem to be slightly more sensitive for route change than others, while commuters also changed route more frequently than people with discretionary purposes.

The respondents, who took another route, were also asked about the length of this alternative compared to the planned route. There seemed to be a relationship between this alternative trip length and the usual trip purpose. Relatively many of the business drivers made a long detour of more than 5 km (30%, against 23% in both other groups). Regarding those making an average detour length of 2-5 km, came the relatively largest part from the respondents with work as trip purpose. And the respondents with discretionary trip purposes made relatively most of the smallest detours. For 13% of them the length of the alternative route was equal to the planned route, while this was the case for 6% of the business people and for 9% of the people with working purposes.

It may be concluded that business drivers are more willing to take an alternative route than others, even if this route is rather a detour way. People with discretionary purposes prefer to stay more on the road where they are, but if they change their route, it results in a route which is not much longer than the planned one. This result was expected on the basis of the higher value of travel time and lower perception of travel costs by business drivers.

This reason might also partially explain why the females are less sensitive to route change: females travel relatively less for business purposes than the males in the survey population.

The impacts on route choices appeared also to be to some extent positively related to the frequency of driving (and inherently, the familiarity with the road network). Among those who drove very frequently (five days a week), 32% of the people appeared to reroute their trips often or regularly, while some 68% only rarely (or never) made a rerouting. Those who drove less than once a week changed route to a lesser extent: 19% often or regular and 81% rarely or never.

No clear correlation was found between the average traffic congestion met (by time of day of driving) and route choice changes. From the people always driving during peak hours the often/regularly route switchers make up 27%, while this share for people only driving during quiet hours is equal (32%). This result might to some extent confirm that benefits from dynamic traffic information are obtained by avoidance of recurrent as well as non-recurrent traffic congestion.

Furthermore, there appeared to be a positive correlation between the average distance of the trip and the length of detours made. From those drivers travelling distances longer than 50 km, 71% make detour distances of more than 2 km, whereas this percentage was substantially lower (53%) for those who travel shorter distances of up to 25 km. Thus it seems that the reluctance to make additional kilometres decreased as the journey length increased. But also some correlation existed here with the trip purpose: those travelling longer distances consisted relatively more of business drivers. Of those travelling more than 50 km, 40% had business motives, while of those travelling less than 25 km, this percentage was only 26%.

Willingness to pay

In addition to the correlation between the willingness to pay for having VMS information available and the gender of the respondents, as described in Section 5.3.7, we have also found interesting correlations between the willingness to pay and some travel features. These are now discussed.

As expected, a correlation was found between the willingness to pay and the usual trip purpose. Respondents with regular business appointments were prepared to pay higher amounts. Approximately 13% would pay more than 40 ECU per annum, against 4% and 1% for commuting and discretionary purposes, respectively. The group with discretionary purposes was very unwilling to pay anything: 62% did not want to pay anything, as against 44% and 51% for business and commuting purposes, respectively.

Generally speaking, business people are more prepared to pay than commuters, who are in turn more prepared to pay than people with other purposes. This is only logical, because the former two groups of drivers have a higher value of travel time than the latter group. The difference between business drivers and commuters may have to do with the fact that travelling to and from work mostly takes place at the same time each day, while business appointments are more likely taking place at varied times on different days. This means that, if the traffic situation is less predictable for business drivers, they might have a larger need for traffic information. Also the fact that in business travel there is a straight transfer payment from employer to information provider, plays an important role here.

The correlation between the willingness to pay and the frequency of driving is not very clear. In general, the frequency with which people make use of the ringroad does not seem to have much influence on the willingness to pay. An exception however, is the group of respondents driving a car over the ringroad one or two days a week. This group is prepared to pay more for the information than people driving more often: 62% is willing to pay something, whereas only about half of the respondents with higher driving frequencies has this attitude. Moreover, the percentage of respondents willing to pay more than 40 ECU is much higher for the driving frequency of 1-2 days a week (15%) than it is for the higher frequencies (5%). However, the total picture shows no clear correlation between these two variables.

Not very surprising was the correlation between the willingness to pay and the availability of route alternatives. Respondents not having an alternative route available were prepared to pay less for road information than respondents having an alternative. However, a considerably high number of 43% of the respondents without route alternatives was willing to pay something. This result re-affirmed that the individual benefits from the VMS information are also obtained by those who had no route alternatives available or do not actively react to the information by rerouting their trip.

The analysis in the present chapter was mainly based on descriptive statistics and contingency table analysis resulting from our field work. In Annex 5A we will present a more sophisticated statistical analysis of the data with respect to route choice behaviour in the RIA system. This analysis is mainly based on Emmerink et al. (1994b).

5.4 Conclusions from the Case Study

The behavioural survey for assessing driver responses to the introduction of road-based dynamic traffic information by means of VMS signs on motorways around the city of Amsterdam has generated quite some interesting conclusions.

The attitudes towards the system were in general very positive, and the impact on route choices of drivers proved to be significant. It appeared that besides a reduction in travel time, drivers may obtain other forms of benefits from the information, like for example a reduction of uncertainty with respect to the traffic situation. Clear differences in attitudes/reactions were found between various segments of car drivers, e.g. between males and females, and between different journey purposes. Some of these were expected on the basis of, for example, differences in the perception of travel time.

Some aspects of the responses indicated that there could also be undesired side-effects, e.g. when alternative routes are (partly) made across secondary roads through the city itself.

A final lesson from this empirical work may be that the commercialization of traffic information among the broad public might be more limited than initially expected - in light of the generally low willingness to pay in the sample -, but also that clear user segments can be distinguished - as identified in the Nested Approach -, among which the market penetration is likely to vary significantly. Such socio-psychological and economic factors will ultimately have important consequences for the future diffusion rates of the new telematics technology. Clearly, the quality of information is decisive for its final market potential.

Annex 5A A Discrete Choice Model for Route Choice Behaviour[1]

Econometric empirical work in a route choice context has generally been based on univariate models. Such models implicitly assume that different types of information have no impact on each other. In reality, this assumption does often not hold true (Emmerink et al. 1994b). Radio traffic information and RIA traffic information both give information on the same traffic situation, and are therefore highly dependent. From a theoretical point of view, any model that investigates the impact of these kinds of information on route choice behaviour should at least allow for dependency. In this Annex, we will extend the generally used univariate models by applying a bivariate ordered probit model. This model allows for a positive or negative correlation in the respective error terms of the underlying response models. If a positive (negative) correlation is found, it implies that an unobserved variable is influencing both endogenous variables in the same (reverse) direction.[2]

The following (endogenous) survey variables are used in our discrete choice model:

Y_1: 'Did radio traffic information (by broadcasting or by RDS) ever affect your route choice in the past three months?'

Possible answers:

1: route choice has never been influenced by radio traffic information;
2: route choice has only a few times been influenced by radio traffic information;
3: route choice has frequently been influenced by radio traffic information,

and

Y_2: 'Did VMS queue information provided by RIA ever affect your route choice in the past three months?'

Possible answers:

1: route choice has never been influenced by RIA traffic information;
2: route choice has only a few times been influenced by RIA traffic information;
3: route choice has frequently been influenced by RIA traffic information.

In order to conduct a useful analysis, these variables were conditioned on having more than one route choice alternative available.

---

[1]The authors wish to thank Richard Emmerink for his assistance in the statistical analysis of this study.

[2]In addition, a bivariate model uses information on two endogenous variables, which generally improves the parameter estimates and decreases the size of the standard errors.

The underlying response model is defined as:

$$Y_1^* = \beta_1 x_1 + u_1 \qquad (5A.1)$$
$$Y_2^* = \beta_2 x_2 + u_2$$

where Y_1^* and Y_2^* are unobserved variables. The error terms u_1 and u_2 are assumed to follow a standard normal distribution, and the correlation coefficient between u_1 and u_2 is given by ρ. Next, we define the constants $\alpha_0^{(i)} < \alpha_1^{(i)} < \ldots < \alpha_m^{(i)}$, where m denotes the number of categories of the dependent variable, i denotes the two endogenous variables, and $\alpha_0^{(i)}$ and $\alpha_m^{(i)}$ are defined as $\alpha_0^{(i)} = -\infty$, $\alpha_m^{(i)} = \infty$. $\alpha_1^{(i)}, \ldots, \alpha_{m-1}^{(i)}$ are estimated with the data. Although Y_1^* and Y_2^* are unobserved, it is known to which category they belong:

$$Y_i = j \ \ if \ \ \alpha_{j-1}^{(i)} \leq Y_i^* < \alpha_j^{(i)} \quad i=1,2 \quad j=1,2,3 \qquad (5A.2)$$

Therefore, $Y_1 = x$ and $Y_2 = y$ if and only if:

$$\alpha_{x-1}^{(1)} \leq Y_1^* < \alpha_x^{(1)} \ \ and \ \ \alpha_{y-1}^{(2)} \leq Y_2^* < \alpha_y^{(2)} \qquad (5A.3)$$

This bivariate ordered probit model was estimated using the maximum likelihood criterion. The results are given in Table 5A.1. Table 5A.1 also gives the estimates of similar univariate models. The reference groups of the dummy variables are shown in Table 5A.2.

Comparing the results of the bivariate model with the estimates of the univariate ordered probit models, we conclude that all the estimates of the explanatory variables are very similar. The parameter estimates of the models reveal that:

- Males are more likely to change routes due to both types of information.
- The exogenous variable 'trip purpose' has a significant effect on the RIA traffic information: People on business trips are more likely to be influenced by the RIA traffic information than commuters.
- The estimates referring to 'trip distance' are somewhat ambiguous.
- Drivers with a flexible arrival time are less inclined to change route due to the RIA traffic information.
- Drivers who are frequently listening to the radio traffic information are more likely to be influenced.[3]
- Drivers who are frequently passing the RIA traffic information sign are more likely to use the information.

[3]The variable 'listening propensity' might in fact be endogenous. However, the estimates of a model in which the listening propensity variable was omitted showed that the inclusion of this variable does not affect the other estimated parameters.

Table 5A.1. Bivariate ordered probit model of route choice adaptations due to radio traffic information and RIA traffic information (t-values in parentheses).

variables	Bivariate radio		Bivariate RIA		Univariate (correlation = 0) radio		Univariate (correlation = 0) RIA	
* constant								
α_1	-0.71**	(-2.37)	-0.53*	(-1.71)	-0.70**	(-2.33)	-0.42	(-1.31)
α_2	1.60***	(5.16)	0.73**	(2.28)	1.62***	(5.23)	0.85***	(2.66)
* age								
< 35	0.15	(0.85)	0.03	(0.19)	0.15	(0.79)	0.03	(0.18)
35 < age < 45	0.18	(0.95)	0.15	(0.82)	0.18	(0.93)	0.14	(0.75)
* gender								
male	0.37**	(1.68)	0.31*	(1.44)	0.39**	(1.74)	0.30*	(1.41)
* trip purpose								
commute	-0.20	(-1.14)	-0.30**	(-1.83)	-0.21	(-1.22)	-0.31**	(-1.87)
business	0.24	(1.20)	0.27*	(1.37)	0.23	(1.13)	0.29*	(1.51)
* trip distance								
< 25 km	0.32	(1.24)	0.14	(0.59)	0.32	(1.27)	0.14	(0.61)
25 - 50 km	-0.38***	(-2.45)	-0.22*	(-1.49)	-0.37***	(-2.41)	-0.21*	(-1.48)
* arrival time flexibility								
flexible	-0.17	(-1.18)	-0.48***	(-3.50)	-0.18	(-1.24)	-0.47***	(-3.44)
* listening propensity								
frequently	0.55***	(4.08)			0.57***	(3.95)		
* passing frequency RIA sign								
> once a week			0.68***	(3.20)			0.81***	(3.60)
once a week			0.73***	(2.47)			0.90***	(2.83)
* correlation								
$\rho(u_1, u_2)$		0.50***	(7.67)			---		

Model: bivariate

309 observations		
-2 Log Likelihood for full model	1006.5	
-2 Log Likelihood for restricted model (estimate α's and ρ)	1079.0	
Likelihood Ratio Chi-square	72.5	(19 d.f.)
-2 Log Likelihood for univariate model (sum of univariate models)	1048.2	
Likelihood Ratio Chi-square	41.7	(1 d.f.)

*: significant at 10%; **: significant at 5%; ***: significant at 1%

Table 5A.2. Reference group of dummy variables.

variables
* age ≥ 45
* gender female
* trip purpose otherwise
* trip distance ≥ 50 km
* arrival time flexibility no flexibility
* listening propensity never and regularly
* passing frequency RIA sign < once a week

The most significant parameter in the results is the correlation coefficient. The estimate of the correlation in the error terms is equal to 0.5. This is extremely high when taking into account the number of exogenous variables used in the model. Also, the increase in the loglikelihood using the bivariate model is very substantial. Consequently, the bivariate model outperforms the two univariate models in explaining route choice behaviour. From the estimate of the correlation coefficient, we conclude that there exists an unobserved variable that has a positive effect on both the route choice influence due to radio traffic information and the route choice influence due to the RIA traffic information. This unobserved variable might be the *propensity to use information*. If an individual has a larger propensity to use information, he or she will be more influenced by both RIA and radio traffic information.

6 Car Drivers' Response and Network Characteristics: An Italian Case Study

6.1 Introduction

Chapter 5 has been devoted to the impact of transport telematics on road users in metropolitan or urban areas. We will in the present chapter focus our attention on inter-urban connections, covering different regions. The main empirical behavioral basis for the investigation of the market potential for advanced transport telematics systems on inter-urban road corridors consists in our study of surveys systematically carried out among different types of actors on the demand side of the market. Such user surveys provide insights into the value added as perceived by the end-user, while they also address the motives for this need for - and adoption of - these advanced technologies, as these exist among different kinds of users. Thus, the key questions to be answered on the basis of these surveys are: which technologies prevail, who are the main users and why - or under which conditions - should they adopt these new technologies? These questions follow the Cascade methodology developed in Chapter 3.

Considering the market development of advanced telematics systems at a pan-European level, there will also be a need to look at the telematics market from a spatial, i.e. regional or site-specific, viewpoint. From this perspective, another relevant question arises: where would telematics have the highest potential, in terms of both expected benefits to the transport system as well as the market among end-users? The impact of telematics on the transport system depends on the nature and size of problems related to safety, environment and efficiency on inter-urban corridors, because clearly road conditions on most European corridors are different in terms of density, congestion, fatality rates, geo-physical conditions, quality of infrastructure etc. The telematics market is thus characterized by spatial heterogeneity, which means that it will be impossible to assess *the* market potential and penetration of *the* advanced transport telematics technology. Instead, various road corridors and/or parts of these corridors may lend themselves for specific applications of such technologies. In Chapter 4 we already have identified representative road corridors and/or segments thereof for which meaningful technologies can be used. In this chapter we will follow up that approach by investigating now also the user side.

The existing differences between corridors or regions in which certain parts of the main road network are located imply that also from the user perspective, several differences may exist in needs for telematics (MARTA 1994c). One may argue, for example, that information on fog will be more desirable for motorway drivers in areas with recurrent fog conditions than in areas where this is not the case, so that such conditions will have implications for the local telematics market among end-users. It is therefore important that research on attitudes towards different forms and types of driver information (in particular with respect to individual telematics equipment) of private motorway users takes place in the context of the features of the (local) road network that is being used. The aim of this chapter is to describe and analyze motorway driver needs with respect to network features that play an explanatory role regarding the relevance, use and adoption rates of transport telematics applications. The analysis will be illustrated by means of a depth investigation of the Italian motorway network.

The structure of this chapter is as follows. In the next section the analysis framework is presented. Section 6.3 contains a functional description of the analysis carried out. In Section 6.4 spatial differences in attitudes of Italian motorway users are described, followed in the subsequent section by a statistical test of some hypotheses that certain attitudes may depend on local traffic conditions. Finally, in Section 6.6 the main conclusions of the present work will be summarized. The Annex 6A describes the software developed in this framework.

6.2 Analysis Framework

The type of analysis used is based on disaggregated data on needs and preferences of motorway users, as they have been revealed in a survey among Italian motorway drivers. These data are linked to various spatial factors concerning features of the road network. First, in a qualitative sense, the spatial variations in attitudes are described. Then, in a quantitative sense, hypotheses are tested about the needs for different kinds of traffic information perceived by the drivers in relation to the various road conditions they meet. To test such hypotheses, a regression analysis is carried out. A possible existing empirical relation has consequences for the transferability of results on drivers' attitudes to other road corridors.

This evaluation approach with a focus on local conditions of networks is supported by using Geographical Information Systems (GIS). A dedicated GIS development platform, called MathGis[1], will be used and extended for the specific analytical problems in our case study: a spatial linkage technique will be developed in order to relate network data and survey results, based on the corridors used by the respondents. Application of this technique enables the investigation of relations as mentioned above and the visualization of spatial differences in transport telematics potential on a more detailed level.

Data about driver attitudes have been gathered from a survey among Italian motorists. From this survey (871 respondents, conducted in October 1993) information was

[1]The development of the MathGis software took quite some time and effort, but had promising results, not only for this case study, but also for other kinds of network related analyses. The Annex 6A will provide some background information about this software. The authors acknowledge the creative contribution made by Maarten Hilferink.

collected about the perceived importance of *types of information* to be offered by telematics. Drivers were interviewed while making a journey. The following data was gathered: location of origin of the trip; location of destination, and perceived importance of information provision for the trip in the following types: route information, information about the trip destination, information about stop & service areas, weather information, congestion information, information about road works, accident information and information about road aid services. Next, it is necessary to assign these revealed attitudes to the Italian road network segments in order to obtain a detailed view on their geographical distribution. Several inferences will finally be drawn on the basis of the spatial analysis undertaken in this chapter.

The subsequent sections in this chapter contain some specific terms related to the analysis that need some further explanation, which will first be given. A series of *functional objects* is used related to both the digitized Italian road network and the survey. They are defined as follows:

- *Point*: coordinate point of a road with a special feature; this could be either a point where the road makes an angle or a node (see below);
- *Node* (subset of *points*): a road junction or a *location* (see below);
- *Location* (subset of *nodes*): origin or destination place of respondents of the survey represented by one single pair of coordinates;
- *Segment*: atomic road connection between two points; these are (atomic) parts of motorways having an average length of about 10-20 km;
- *Link:* path of connected segments between two nodes;
- *Route:* path of connected links between two locations;
- *Set of Alternative Routes (SAR):* set of feasible route alternatives between the two locations (each survey respondent is assigned to a SAR; a detailed description follows in Section 6.3);
- *Tollgate:* A node which is also an entrance/exit point of the autostrada network.

After this presentation of some main ingredients of our analytical framework, we will move to a presentation of the functional structure of our spatial analysis in the next section.

6.3 Functional Description of the Spatial Analysis

The core of our analysis is the assignment of the attitudes towards different types of information provision, as revealed in the driver survey, to the road network segments. Each segment is assigned a certain value on the basis of the revealed attitudes of the respondents that could pass that particular road segment. Before this assignment may be carried out, two problems have to be solved. First, for each respondent, a set of feasible routes has to be generated. Second, for each road segment, a weighting factor for each assigned respondent has to be determined. Figure 6.1 shows the relationships between the entities discussed and analyzed. The steps to be undertaken will now be described in more detail.

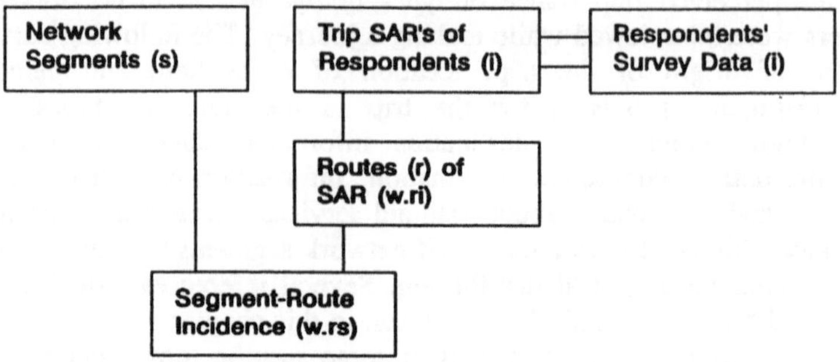

Fig. 6.1. Relations between entities.

Step 1: Choice Set Generation

On the basis of the known origin and destination points of motorists, for each respondent a set of feasible routes has to be generated. Shortest (or fastest) path searching is a standard GIS capability; however, just calculating the shortest route was not sufficient in this case. Car drivers usually have several route options to choose from, so that there is always the possibility that a respondent has chosen not to take the shortest route. Furthermore, the revealed attitudes are considered to be related to a choice set of routes, containing the shortest/cheapest path as well as some realistic alternatives.

In order to facilitate working more easily with route choice sets, we have employed a separate spatial entity, namely a SAR (see Section 6.2). A SAR is defined as a set of feasible routes from a fixed origin to a fixed destination. The feasibility criterion is implemented as follows: a route is feasible if and only if each road segment on the route brings the driver *closer* to his or her destination and *further away* from his or her origin, where *closer* and *further away* are measured by the shortest distance through the road network.

Step 2: Weighted assignment of survey data to segments

Linkage weighting factors are defined in order to support the aggregation towards road segments. The following defintions are used:

w_{ri}: the weight of route r in SAR i to which route r belongs, is defined according to the relative length of a route: M_r/(sum M_r), where M_r is a variable indicating the attractiveness of a route, relative to the most attractive route for the SAR. Currently, M_r is defined as $\exp(-\text{ß}*[\log(Length_r)-\log(MinLength_i)])$ with ß = 2, which equals to $(MinLength_i/Length_r)^2$;

w_{rs}: the weight of route r for determining attributes of network segment s is defined as M_{rs}/(sum M_{rs}), with M_{rs} defined as $Length_s/MinLength_i$.

Each quantitative property found in the survey can now be assigned to network segments s according to the relationship P_s = sum (w_{rs} * w_{ri} * P_i). The weighting mechanism ensures that, although the influence of each survey respondent is differently spread over the road network, the summed influence of each respondent is equal, since the sum of M_{rs} * w_{ri} for a SAR i always equals 1.

6.4 Description of Observed Spatial Variations in User Attitudes

After applying the successive steps of Section 6.3, various maps of average user attitudes towards different types of information provision have been plotted. Three of the eight information types were sufficiently significant and were selected for further analysis: weather information, information about congestion, and information about accidents. The assignment of these revealed preference to the network are represented in the maps of Figures 6.2, 6.3 and 6.4. In order to interpret the spatial patterns shown in these Figures in a proper way, it must be noted that statistical biases in the sample cause, for example, a generally low perceived importance on Transalpine road links, since the survey included only a few respondents travelling outside Italy.

From Figure 6.2 we can derive that *congestion information* is perceived as highly important in some very specific parts of the Italian motorway network:

1. the orbital ring around Rome;
2. the corridors Turin-Modena-Florence-Rome, Venice-Bologna, Rome-Pescara and Naples-Foggia.

The relatively moderate and low preference for congestion information on the Milan orbital ring may of course be due to a statistical bias in the sample, but it is likely that this is caused by the fact that the Milan ringroad is continuously (throughout the day) heavily congested, so that such information does not add very much to the needs of the motorists.

Weather information (Figure 6.3) seems to be the more desired type of information that Italian drivers would like to receive. The unequal spatial distribution of this preference is quite clear. In the Southern regions we observe that - except for some small, isolated sites in the network - the perceived importance assigned to weather information is generally low and moderate. In contrast to this, in central and Northern Italy, especially in the North-Eastern part, we observe that the perceived importance attached to this class is generally high or at least moderate. The preference for weather information seems to be strongly connected with the preference for accident information, especially on the Venice-Brescia-Piacenza and on the Parma-La Spezia corridors. The high preference for dynamic weather information on the Venice-Brescia-Piacenza corridor can likely be explained by the recurrent foggy weather conditions in that area, while in the mountainous area between Parma and La Spezia the influence of severe weather conditions may be aggrevated by the latitudinal road conditions.

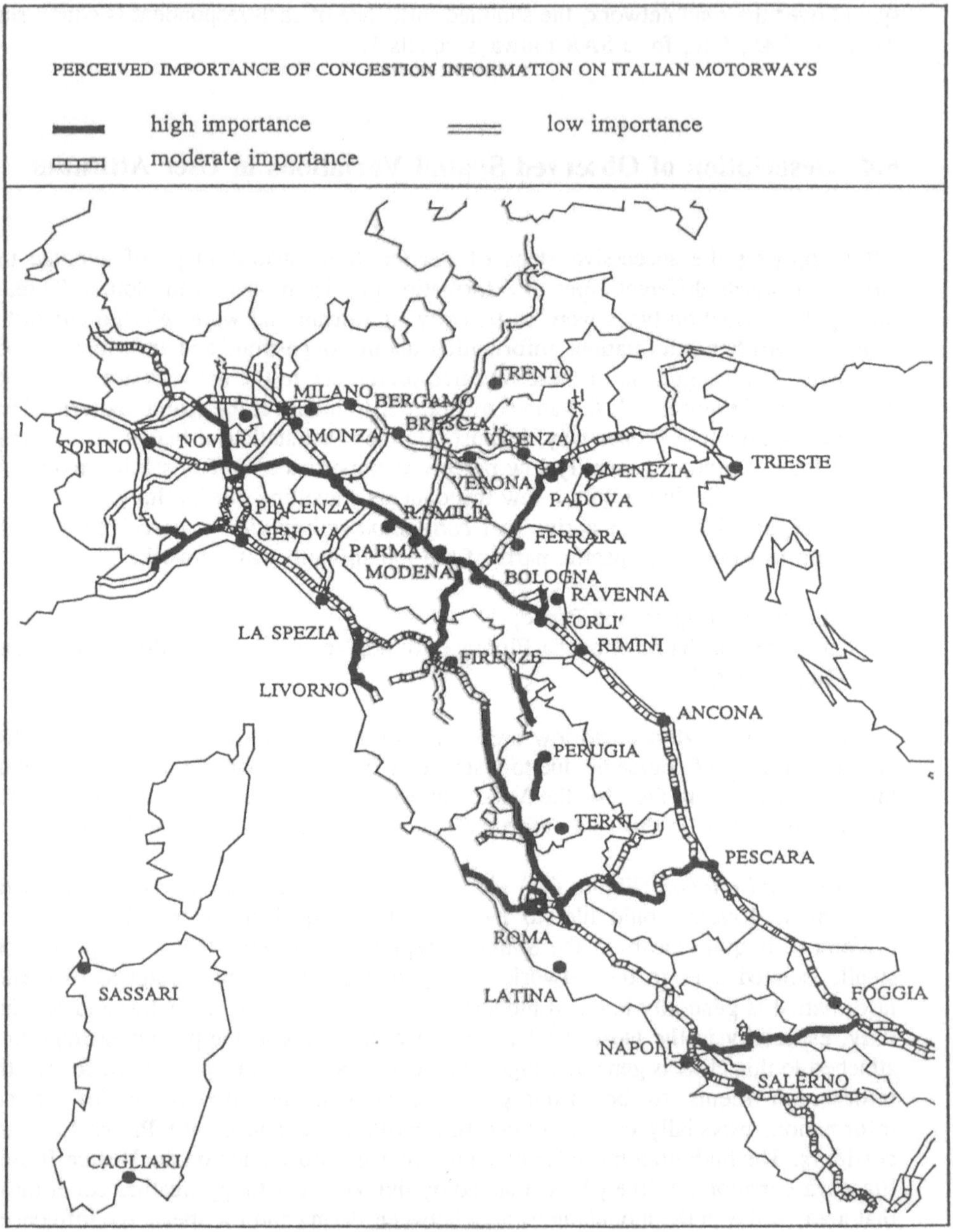

Fig. 6.2. Perceived importance of congestion information Italian motorways.

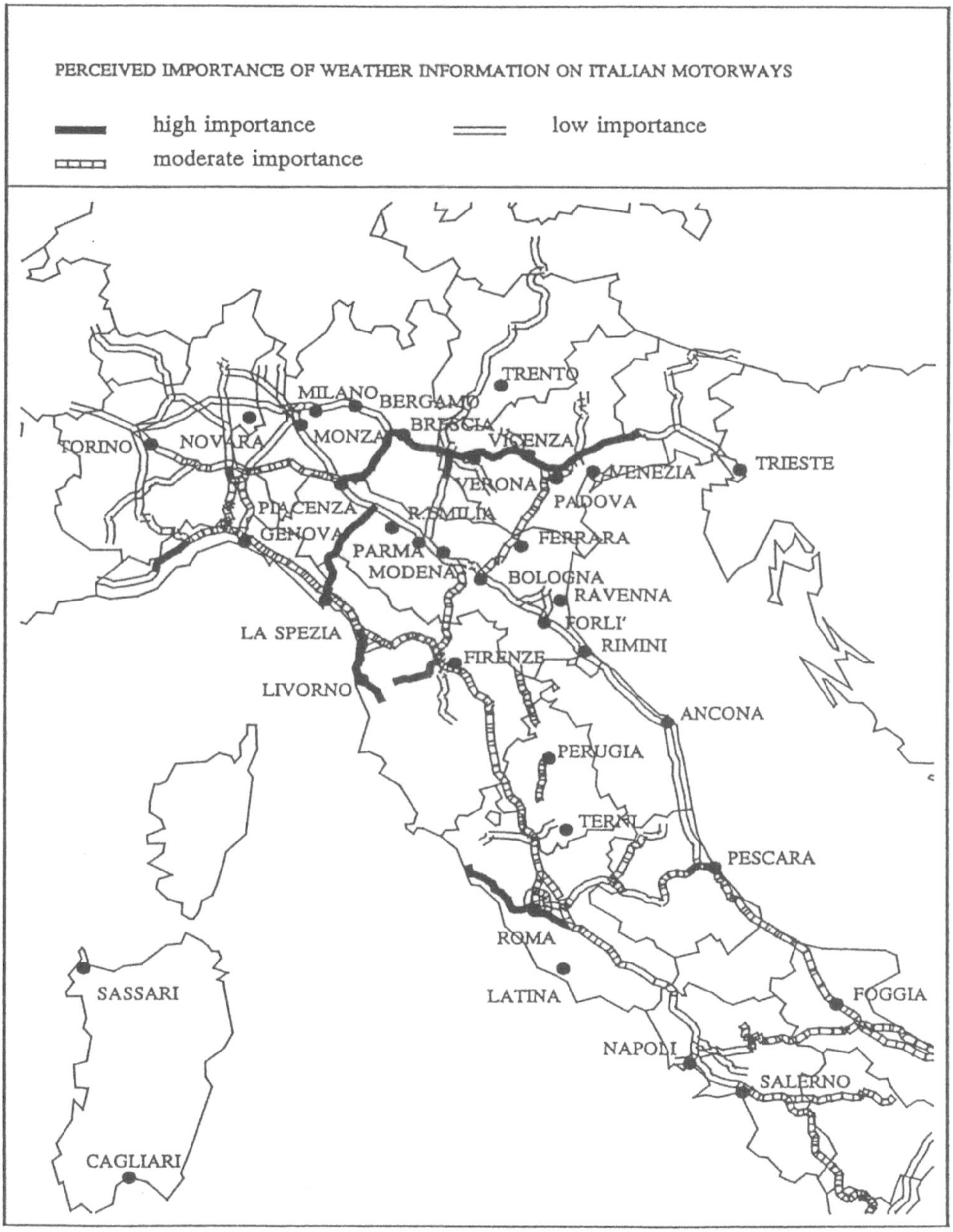

Fig. 6.3. Perceived importance of weather information on Italian motorways.

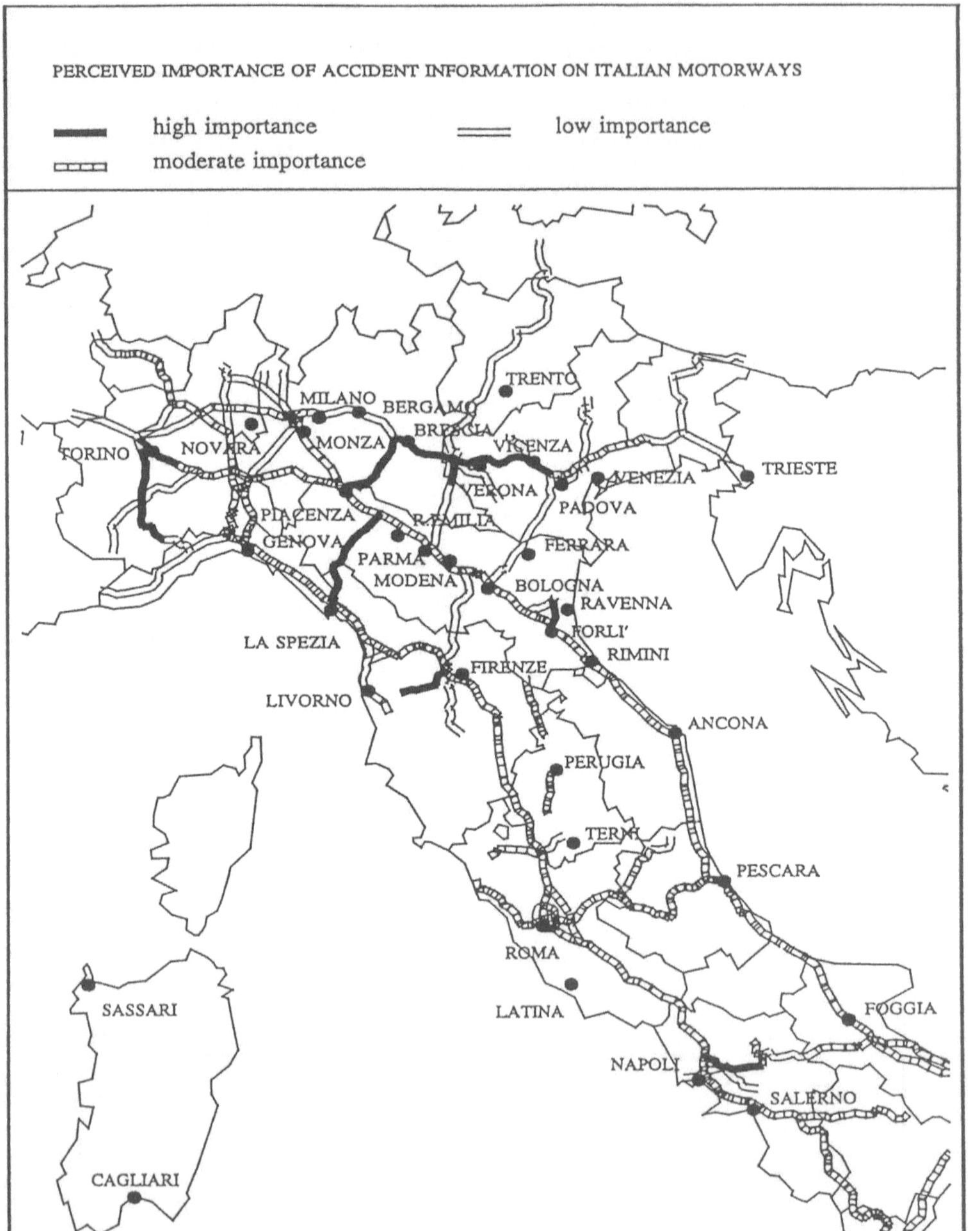

Fig. 6.4. Perceived importance of accident information on Italian motorways.

The provision of *accident information* (Figure 6.4) seems to play an equally important role on these road network links as weather information. Such a role is particularly clear for the risky roads through mountainous areas in the region between Florence, Livorno, La Spezia and Parma. On the contrary, there is a low value attached in the sample to information on accidents on the Bologna-Padua-Venice-Trieste corridor, while it is generally acknowledged as also this latter corridor is a relatively risky corridor due to the nightlife in this surrounding coast area. This is somewhat unexpected, but here it may be the case that all frequent road users know the risk on this corridor and do not derive a higher utility from the provision of such information. These results are summarized in Table 6.1.

Table 6.1. Perceived importance of traffic information in Italy: comparison between regions by means of map analysis.

Region	*Congestion information*	*Weather information*	*Accident information*
Northern	generally moderate to high	generally moderate to high; in particular on links with recurrent bad weather conditions and in mountainous areas	generally moderate to high; in particular on links with recurrent bad weather conditions and in mountainous areas
central	generally moderate to high	generally low to moderate; high only on risky (mountainous) links	generally moderate
Southern	generally moderate	generally low to moderate	generally moderate

6.5 Analysis of Influence of Congestion and Road Accidents on Drivers' Attitudes

In this section a regression model is used in order to assess the role of some key features related to the Italian motorway system towards the driver's perceived needs of certain types of transport telematics information. In particular, the following hypotheses are considered:

1. The perceived need for dynamic congestion information is higher when the average level of congestion the driver (is used to) meets during his car trip is higher.
2. The perceived need for accident information is higher when accident rates are higher.

We will now try to explain here the perceived need for information, which is principle a variable which indicates the order of magnitude of the perceived need. Ideally, this type of information should be analyzed by means of ordered discrete choice models, but in case of a high number (5 or more) of levels, a simple regression analysis may be more satisfactory, as it will unlikely bias the results (based on the potential of ordered discrete choice model statistics).

In order to identify also the importance of large regional differences, the location, expressed as a two-dimensional variable, has been included into the regression. Results of these regressions are presented in Table 6.2. The variables used to test these hypotheses are:

- Perceived importance of information on congestion (CON), scaled from 1 (high perceived need) to 5 (low perceived need);
- Perceived level of importance of weather information (WEA), scaled from 1 (high perceived need) to 5 (low perceived need);
- Perceived level of importance of accident information (ACC), scaled from 1 (high perceived need) to 5 (low perceived need);
- Index for the average congestion level or Level of Service (LOS): a description of the unit used for the LOS can be found in Chapter 4; it varies between 0 (low average congestion) and 1 (high average congestion);
- Index for accident rates per kilometre per constant number of vehicles (ACCRATE) (which is taken as a measure of individual risk);
- Longitude (X-coordinate) of road segments (X-COORD);
- Latitude (Y-coordinate) of road segments (Y-COORD).

Table 6.2. Results of the analysis

Considered exogenous motorway attributes are: LOS, ACCRATE, X-COORD and Y-COORD.
Dependent variable CON
F = 2.43260 Signif F = .0462

Variable	B	SE B	Beta	T	Sig T
LOS	-.109003	.090165	-.045058	-1.209	.2271
ACCRATE	-.002541	.000972	-.099584	-2.613	.0092
X-COORD	-.000232	.000150	-.104323	-1.542	.1235
Y-COORD	-.000115	.000144	-.054054	-.803	.4221
(Constant)	3.054163	.080389		37.992	.0000

Log-likelihood Function = -461.074971

Dependent variable WEA
F = 11.43394 Signif F = .0000

Variable	B	SE B	Beta	T	Sig T
LOS	.373248	.103831	.130792	3.595	.0003
ACCRATE	-.003049	.001120	-.101282	-2.722	.0066
X-COORD	-.000605	.000173	-.230814	-3.495	.0005
Y-COORD	-.000150	.000165	-.059531	-.906	.3652
(Constant)	2.882890	.092573		31.142	.0000

Log-likelihood Function = -563.386894

Dependent variable WEA
F = 17.90228 Signif F = .0000

Variable	B	SE B	Beta	T	Sig T
ACCRATE	.004065	.000864	.173997	4.704	.0000
X-COORD	-.000550	.000134	-.270467	-4.120	.0000
Y-COORD	-.000544	.000127	-.278135	-4.268	.0000
(Constant)	3.352497	.054244		61.804	.0000

Log-likelihood Function = -375.870497

Interpretations
We can see from Table 6.2 that the expected relationship between actual congestion levels and the need for dynamic information on congestion cannot be statistically confirmed. Yet this need for congestion information as well as information on road accidents is statistically significantly higher in the Northern part of Italy when compared with the Southern part.

It appears also that our second hypothesis about the relationship between accident rates and attitudes toward accident information is not confirmed at all. In fact, roads with higher accident rates appear to have drivers who have a negative interest in accident information, although on a regional level the relationships are positive (in the Northern part of Italy, accident rates are higher than in the Southern part and needs for accident information are also higher).

Our analysis comes thus to the intriguing conclusion that road users on Italian motorways are likely less to be interested in accident information on high risk roads because of psychological reasons, such as:

- no interest in discomforting messages;
- higher valuation of messages that are less likely to occur; the content of information in an information theoretical sense is said to be higher when the event it represents was less likely to occur (Theil, 1967). The more *unlikely* the event is before the message on its realization, the larger is the information content of this message;
- saturation of risk awareness.

This seemingly counter-intuitive result shows that relationships between network characteristics and attitudes towards information provision are less straightforward than expected and that driver attitudes might not directly be determined by the usefulness of information provision.

6.6 Conclusions

Evidence from a survey among Italian road users supports the following conclusions:

1. Congestion information devices seem to have more potential in the Northern part of Italy. However, a significant statistical relationship between actual congestion levels and the need for dynamic information on congestion has not been found.
2. Weather information services seem to have more potential on those road links located in areas which are subject to frequent severe driving conditions such as foggy weather circumstances.
3. It seems that accident signalling technology has a greater potential on corridors located in mountainous areas. However, our analysis showed that on the individual level, drivers seem to be less interested in information provision about accidents on links with higher accident rates.

The Italian case study has brought to light that various factors affect driver attitudes towards information providion and that knowledge of these factors can help to identify potential needs and (as shown on the basis of the attitudes towards accident information provision) can uncover potential risks when implementing transport telematics.

Relationships between the subjective needs for information provision and the objective network characteristics, as investigated in this chapter, can identify which factors determine informational needs. In this way, the locations and circumstances under which information needs occur can be identified on the basis of network characteristics, which can contribute to a more effective usage of telematics technology on the road.

Summarizing, we observe that this disaggregate analysis based on a single case area performs quite well in terms of general needs assessment. Clearly, only a more comprehensive behavioural model fed with individual survey data may provide more precise policy recommendations. For the time being, it is evident that *the* market for transport telematics services does not exist. There is a great variety of local or site-specific conditions that is decisive for the market potential of telematics. Furthermore, there is also a great variety of user groups. In conclusion, a segment analysis based on the Nested Approach is a necessary analytical framework for assessing the potential and use of transport telematics.

Annex 6A Software Description[2]

In order to perform the computational steps as required for the linkage of survey respondents to network links, one has to use a sufficiently powerful platform for implementation. The need to assign spatial objects to other spatial objects is found in almost any field of spatial analysis and one would expect to find assignment operations in any analytical GIS software package. However, the specific algorithmic structure of network assignment problems requires an original and new implementation, which is certainly the case for linking respondents to road segments (see also Hilferink and Rinzema 1995). Therefore, it is no surprise to find no standard GIS software packages with support for our analysis.

What is required for the implementation of a specific spatial analysis, such as linking respondents to road links/segments, is an application framework with a re-usable code. Current object-oriented programming (OOP) technology enables the development of such application frameworks. Although many general purpose application frameworks for Graphical User Interfaces (GUIs) and data-access have been developed, it was not possible to use a Geographic Application Framework that meets our requirements. Therefore, it was necessary to develop a new approach named *MathGis*. At the start of the development of *MathGis*, the following demands were formulated:

- The framework should contain class definitions for all objects commonly found in a GIS, such as points, arcs, polygons, networks, routes, SAR's, layers, user-attributes, coordinate systems, map windows, table windows, selections, classification diagrams and drawing styles.
- A script language should be used in order to define an entire application by the objects which constitute it; the same script language should be used to save and reload the current state of all application objects.
- Classes representing the application and result of specific calculations should be fed into the framework
- Topological relations between entities should be expressed using available objects.
- Changes to the state of any object should result in an automatic update of all dependent objects.
- Time expensive calculations and drawing operations should be postponed when a user triggers new commands in order to increase the speed of interaction with the user.

In an object oriented application framework, the application developer can use its components, either by only using existing classes, or by deriving new classes from existing ones in order to implement application-specific behaviour. In either way, the code is re-used and the standard operations or methods remain available.

---

[2]The authors wish to thank Maarten Hilferink for his assistance in the GIS work of this study.

The GIS application framework *MathGis* contains classes of objects found in any Geographic Information System. The difference with other GIS packages is that these classes are available for the application developer for the creation of new derived classes with a problem-specific behaviour.

Within MathGis, a division is made between primary data and secondary data. Primary data, encapsulated by primary data classes, describe the problem related objects, such as roads, toll-gates, respondents, etc. Secondary data, encapsulated by secondary data classes describe the application components, models, interface elements and their interrelations.

Primary data classes

Primary data classes are sub-divided into elementary (geographic) data classes and composite data containers. The following base classes are available for the definition of primary data objects. Derivations can be made in order to implement objects with more specific behaviour. The different attributes of problem related objects are stored in separate composite data containers.

Geographic data classes:

TGeoPoint defines objects with a geographic location, such as nodes of a (road) network or centroids of (administrative) areas

TGeoArc defines a collection of *TGeoPoint* objects, used for arcs, polygons, etc.

TGeoRect defines geographic rectangles used for the implementation of rectangular indices for speeding up spatial selections of geographic data. All previous geographic data classes are able to supply their rectangular extents.

TLink defines a link between two nodes, represented by TGeoPoint objects.

Data collection classes:

TAbstrArr<type> serves as an abstract container for objects of class *type*. It has member functions to get and put elements from and to the container.

TmemArr<type>, derived from *TabstrArr<type>*, implements storage in memory arrays. All elements can be loaded and saved to a stream file.

TVirtArr<type>, derived from *TabstrArr<type>*, serves as a base class for virtual arrays who calculates each element at the moment it is requested for. Derivatives from this class can be used to define attributes as a function of other attributes without the need to allocate resources for their storage.

TNetwork defines a collection of TLinks objects and maintains the topological relation from nodes back to the referring links.

TRoute defines a minimal path tree in a network.

TSAR defines a SAR between an origin and a destination node, containing all routes that have only links that bring a road user away from the origin and more nearby the destination.

Secondary data classes

Secondary data classes describe the application components, models, interface elements and their interrelations. Secondary data objects can use other secondary data objects for input. With this mechanism, model components can be plugged into other models dynamically at run-time.

Interrelations between secondary data objects must be managed in order to determine what should be done when some information is changed somewhere. Therefore, a secondary data object can have registered suppliers (which are used as input) and consumers (who use it as input).

A secondary data object can be made invalid, which will recursively invalidate its consumers. Invalidated objects can be regenerated which only takes place after all suppliers are regenerated. With this mechanism, integrity of all in-between results is maintained after changes are made in any model specification or primary data input. Regeneration takes place automatically when no user events are to be processed and is suspended when new events are received.

All secondary data objects have a name and can draw this name into a ListBox (which is a standard object in all Graphical User Interfaces).

Secondary data classes are sub-divided into data description classes and desktop description classes. The following base classes are available for the definition of secondary data objects. Derivations can be made in order to implement objects with more specific behaviour.

Data description classes:

TWorkSpace defines workspace objects. A workspace identifies a directory containing related data. Each workspace contains entities.

TEntity objects represent entities known by the application; entities are a collection of imaginary data objects who have the same attributes defined. Entities grant a framework for defining relations between data objects. Each entity has attributes. Each entity belongs to a workspace. Each entity has a selected attribute and a highlighted attribute, determining which elements are selected and highlighted.

TAttribute objects represent attributes to entities. Attributes have a domain which specifies the type of values they contain. Attributes can be anything related to an entity, which is more than what could be represented as columns in a table.

Classes, derived from *TAttribute*, are TKeyIndex, TRefIndex, TRelIndex and TSelIndex, used for sorting, referencing, relating and selecting members of entities. Network topology, required for network path calculations, can be maintained using these derived classes.

TGisRect objects maintain rectangular indices on geographic attributes with the purpose to accelerate geographic selection and visualisation.

TDomain objects represent domains which determine the values of attributes.

Desktop description classes:

TCoordSysBase serves as a basis of coordinate system objects used in geographic views.

TMapView objects are windows containing a geographic map. A map has several layers and a coordinate system.

TTableView objects are windows containing a table of text-attributes of the same entity.

TGisLayer objects are layers in a geographic view.
TDiagram objects represent classification diagrams. Classification diagrams are used to group, rank, or visualize data attributes. A classification diagram contains several class descriptions.
TClassDescr objects represent the classes defined in a classification diagram.
Each classdescr refers to a global *TStyle* object, of which three sub-classes exist: *TPenStyle*, *TGroupStyle* and *TRankStyle*.

7 Public Transport Information Systems: An English Case Study[1]

7.1 Introduction

One of the key problems for passengers using urban bus services is its unreliability, mainly caused by congestion in the system. Real time information allows passengers to make decisions based on the actual arrival time of the bus, rather than the expected time. If levels of satisfaction with the bus service are raised, then the improvement in the quality of that service will result in a greater demand from existing bus users. It may even result in car users and others switching to the bus service. Recently in Europe some interesting case studies have been carried out which aim to assess behavioral responses of end-users to transport telematics applications (Argyrakos et al. 1994). Two of the case studies carried out in conjunction with the EU DRIVE II programme of field trials which have made use of a range of public transport information systems are presented in this chapter:

1. In *Birmingham* (UK), a public transport traveller information system has been tested for buses. The main objective of this trial was to create conditions in which people would increasingly choose public transport for their journeys, especially for journeys to work. This trial consists of a hotline enquiry service, interactive in-street terminals, real time countdown displays at bus stops along a single corridor and interactive in-home terminals. These systems provide users with information on timetables, fares, optimum routes, route maps, disruptions and special services. Some of the findings from the real time information systems are reported here (Section 7.2). 'Before' and 'After' surveys have been carried out in the south of the city to assess user response to the real time information boards installed at five bus stops.

---

[1]The material for this chapter has come from the EU DRIVE II project BATT, and we are grateful to all partners in the consortium for their contributions - TRENDS (Athens), FUA (Amsterdam), TTR (Oxford), and UCL (London). We are also grateful to the ROMANSE (Southampton) and QUARTET (Birmingham) projects for their help in collecting the data, and for support from Hampshire County Council and from CENTRO (Birmingham).

2. In *Southampton* (UK), bus services are provided by six operators on 56 routes, although in the deregulated environment this is likely to change at any time. After a period of considerable uncertainty following deregulation in 1986, the network is now relatively stable, yet one of the key concerns of passengers is the unpredictability of road based public transport systems. Real time information systems can provide passengers with some reassurance of when their buses are expected to arrive at the stop, allowing them to decide whether to wait for them or to make use of the expected waiting more effectively, or even switching mode if the wait is long and they are time-constrained.

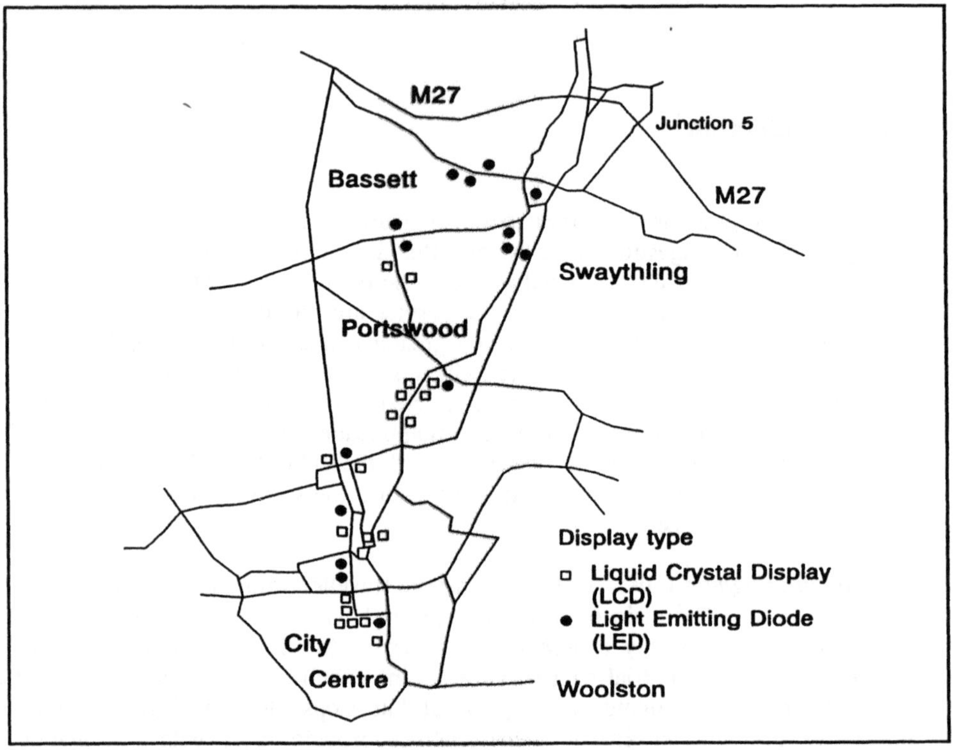

Fig. 7.1. The STOPWATCH corridor.
Source: ROMANSE (DRIVE II).

A real time system (STOPWATCH) has been installed in Southampton and has been fully operational since January 1994. The trial corridor is seven kilometres long and features twelve different routes which are provided by the major operators (see the map in Figure 7.1). The vehicles are equipped with in-bus computer units (IBCU) and receive location information from roadside beacons strategically placed both on the periphery and at key locations within the corridor. This information is updated by distance data transmitted from the bus odometer. The bus location is identified on a map on the computer screen and the data of current and historical journey time is used to calculate the arrival time of the bus. This information is then transmitted to the relevant displays via a radio paging system. A total of 44 bus stops are fitted with LED and LCD displays while 114 vehicles are fully fitted with the in-bus computer units. Both three-line displays show the service numbers of up to the nearest five buses together with their destination and approximate arrival time in minutes. If required, special messages can be sent to any or all bus stops and these are displayed on the bottom line of the signs which can be scrolled.

The presentation of the data from the two case studies is contrasting, as the scale of the implementation of the technology varies and as the timing of its introduction varies. The Southampton information system has been running for a longer period, so that responses are more likely to represent a stable situation, and the number and scale of the application is much greater than that in Birmingham. Similarly, the size of the 'Before' and After' samples is much greater. Consequently, a much greater level of detail within the data can be extracted and the balance of this chapter reflects these differences.

7.2 Results from Real Time Passenger Information Systems in Birmingham[2]

7.2.1 'After' Results

The sample of bus users interviewed were heavily dependent on public transport services as only a half of respondents lived in a car owning household, and most did not have a driving licence. Most were regular bus users, making at least one trip on the bus every day, and over a half of the sample made use of the bus on four or more days a week. However, as with many frequent users, under 20% knew the timetable, and over three-quarters of the bus users did not set out to catch a particular bus. Overall, there was a view that the buses tended to run on time along the corridor. This profile of users is typical of that in many European cities, so that the responses to the real time information systems is likely to represent all bus users.

The introduction of real time passenger information displays at bus stops has been met with a positive response. The technology was particularly positively viewed by the principal users of the services, namely the female travellers, the elderly and the young - over three quarters of the sample considered the displays to be 'useful' or 'very useful'. The information was considered to be of particular use to the regular users of the bus services, and the readability and location of the signs were again seen in a positive perspective. Two thirds of the sample perceived the real time information displays to be more accurate than the printed timetable information.

In short, although this was only a small scale implementation at five stops along one corridor in Birmingham, the results were very positive from all groups, particularly

---

[2]Much of the material used here comes from EU DRIVE II BATT Deliverable 11-B, The Impact of Providing Real-Time Passenger Information at Bus Stops: Results of the Birmingham 'After' Survey and Comparison with the 'Before' Survey, prepared by Transport and Travel Research, Oxford.

those who are regular users of the bus and depend upon the bus service for their basic transport. We now turn to the differences and changes in behaviour resulting from the introduction of real time information systems.

7.2.2 'Before' and 'After' Analysis

The introduction of real time information systems does not seem to have generated more use of buses by new users. For 94% of the sample, the information has had no impact of the numbers of journeys made. In the 'Before' survey, some 25% of the sample stated that the introduction of real time information would result in an increase in the numbers of journeys made by bus. However, the actual results from the 'After' survey suggest that the real number is only 1%. There has been a massive overestimation of the impact of real time information of travel behaviour.

The introduction of real time information does not seem to have led to improvements in the perceptions of bus reliability. Two thirds of those who had experience of using the bus service before the introduction of real time information did not perceive any change in the reliability of bus services since the introduction of the technology. However, the information systems does seem to have encouraged a greater sense of adventure among bus users. Travellers have used the real time information to make better use of their time whilst waiting for the arrival of the bus by going shopping and carrying out other business.

7.2.3 Conclusions

The results from the small scale field trial in Birmingham are disappointing in terms of the potential trip generation effects of the real time information system. But it should be noted that the implementation of the technology is recent and the longer effects may be greater. The behavioral responses are much more subtle. The real benefits have been in the greater flexibility and the feeling of control over time. High quality information has allowed alternative activities to be carried out and a greater satisfaction with the service itself. Passengers have tended to turn up at random at the bus stop and the information gives them reassurance that a bus is coming (or not) and when.

This greater flexibility has not resulted in new travellers to the bus service, nor has it resulted in a further shift of passengers away from the bus. This may in part be because the sample interviewed was very much dependent on the bus service - they had no alternative mode available. There has been no evidence of new travellers on the buses (first time users). Again, this is disappointing as one of the main justifications for real time information systems is that new trips will be encouraged by existing and first time users as the quality of the service has been improved. If this has not taken place in Birmingham, then there may need to be a promotion campaign to inform users of the information systems, together with the other schemes in the city (BATT, 1995b).

Stated preference surveys in other UK cities (e.g. London and Liverpool) have suggested that people would be prepared to pay more for high quality real time information systems, particularly if there are also journey time savings. These can in turn be translated into patronage increases and additional revenue for the operator.

However, detailed monitoring of schemes in several cities, including Birmingham, has not been able to detect significant increases in patronage. In the longer term trends in patronage decline may be arrested, but increases in patronage are unlikely. As we shall see in the second case study in Southampton (Section 7.3), some increase in patronage from a variety of sources has been identified.

The conclusion reached from the behavioural analysis of user response to the introduction of real time information systems in Birmingham has to be that they promote a 'feel good' factor among existing bus users, but even this may only have a novelty value. Without any measurable impact on bus patronage, it is difficult to justify the levels of investment associated with real time information systems. The installation of real time displays forms one part of a comprehensive package of measures to improve the quality of public transport services in Birmingham, but it may take a substantially greater time to see the full impacts. The surveys here were carried out only a month after the real time information displays were introduced and tested (for accuracy and reliability of information).

7.3 Results from Real Time Passenger Information Systems (STOPWATCH) in Southampton[3]

7.3.1 Aggregate Analysis of the 'Before' and 'After' Surveys

The aggregate analysis of frequencies has been carried out to establish whether the 'Before' and 'After' samples in Southampton are comparable (Table 7.1). Although the two data sets were similar in most respects, significant differences were found in the age distribution and the employment status. These differences can be explained by the timing of the two surveys. The 'Before' surveys were carried out in Summer 1993 and the 'After' surveys in Autumn 1994. In the 'Before' surveys, the composition of the sample by trip purpose (and consequently by age and employment) meant that not many trips were recorded for education purposes, and there were a higher proportion of elderly people. In the 'After' surveys there were more education trips as well as a greater proportion of younger respondents (Tables 7.1 and 7.2). These two tables give a clear indication of the differences between the two samples and the main differences between the unweighted samples in terms of the STOPWATCH effects. This has necessitated the introduction of a weighting procedure between the two samples (BATT 1995a).

No changes have been made to the 'After' sample, but the 'Before' survey data have been weighted to compensate for the lower than expected number of young people (17-24 age group) in the sample and the higher number of elderly people (over 60) in the sample. Changes have also been made to reflect the under representation in the student categories, and the over representation in the full employment and retired categories.

---

[3]Much of the material presented here forms part of EU DRIVE II BATT Deliverable 11-A, Results of Southampton 'After' Surveys and Comparison with the 'Before' Surveys, prepared by the Bartlett School of Planning, University College London.

All of the analysis of the 'Before' and 'After' survey data has been made with the weighted 'Before' data (Sections 7.3.3 and 7.3.4).

Table 7.1. Summary of differences in 'Before' and 'After' surveys.

Variable	'Before'	'After'	Significance of results
Gender	Same	Same	Not significant
Age	More elderly	More young (17-24)	Significant at the 90% level
Employment	More employed full time and retired	More full time students	Significant at the 99% level
Car ownership	Same	Same	Not significant
Public transport	More irregular users	More regular users	Not significant
Car use	More who never use the car	More occasional car users	Not significant
Trip purpose	More work trips More leisure trips More visit trips	More school trips More shopping trips	Significant at the 99% level

Table 7.2. STOPWATCH effect.

Variable	'Before'	'After'
Trip generation effect	16% intention	4% actual
Corridor effect		76% of the 'After' sample had used bus services in corridor before STOPWATCH
Service information source	-	22% electronic displays
	47%	50% experience/previous knowledge
	23%	12% timetable at bus stop
	24%	12% timetable booklet
Mode and route change effect (13% of 'After' sample)		38% walk all way 30% walk to another stop 13% shop/bank back to original stop 5% shop/bank walk to another stop 7% hailed taxi/lift 7% something else

7.3.2 Disaggregate Analysis of the 'After' Surveys

In the 'After' survey, there were some limited differences observed by gender, by trip purpose, car availability, employment and use of the bus route prior to the STOPWATCH system being introduced, but it was age and employment where most differences were found (see Table 7.3). The young were the most frequent users of the bus and had a positive view of the service in terms of its quality and levels of satisfaction. They also made new trips on the route and preferred the electronic displays, but were more likely to seek alternatives if the expected delay was lengthy. The elderly almost mirrored this perspective as they were intermediate users of the bus route, but had a much more traditional view of the service quality and the usefulness of the electronic displays. They were more tolerant of late buses and had a generally lower level of expectation. But all users stressed the importance of reliability and frequency as being the two crucial elements of a high quality bus service.

Table 7.3. Sample breakdown by age.

Young	Elderly
• Most frequent bus users	• Intermediate level of bus users
• New users on route	• Existing users of the routes
• Positive view on quality of public transport	• Importance of reliability and frequency
• Higher levels of satisfaction with the service	• Lower expectations of bus services
• Preferred electronic displays	• Preferred printed timetables
• More likely to leave the bus stop if a long delay is expected	• More tolerant of late buses
• Higher levels of car availability and use	• Lower levels of car availability and use

The sample breakdown by employment (Table 7.4) was slightly more complex with three clear groups - those in full time employment or part time employment; housewives, unemployed and students; and the retired. It was the students who gained most from the STOPWATCH information, whilst the other groups used traditional forms of timetable information. Those in employment mainly used their experience, but also had high levels of knowledge on STOPWATCH. The retired again relied on experience and the printed timetable, and had low levels of knowledge of STOPWATCH.

In terms of the analysis by public transport use (Table 7.5), trip purpose and car use, there were again clear differences between various categories of respondents. Frequent bus users (those who used the bus on 4 or more days a week) perceived services to have higher frequencies with relatively short waiting times, but they set out for a particular service. They made greater use of experience and had a high knowledge of the STOPWATCH system, but they placed a low importance on the electronic display. However, they had made a limited increase in their use of the bus as a result of STOPWATCH, and were more likely to find alternatives if their waiting time is high. Infrequent bus users on the other hand were the reverse of the frequent users and had a generally lower level of satisfaction with the bus service. However, they were more likely to use it, and they gave a generally positive view of the electronic display.

These conclusions were reinforced by the trip purpose analysis (see Table 7.6) as frequent bus users were mainly using the bus for work and school purposes, whilst the infrequent bus users were visiting friends, shopping or carrying out leisure trips. The analysis by car use (see Table 7.7) was also consistent, as the frequent car users were the infrequent users of the bus. This group was not able to judge the accuracy of the electronic information system, but wanted an increase in bus reliability. They were not likely to leave the bus stop.

Table 7.4. Sample breakdown by employment.

Full time or part time employment	Housewives, unemployed, or students	Retired
• Set out for a particular bus	• Wait for a bus to turn up	• Set out for a particular bus
• Information from experience	• Information from electronic display (students) the timetable booklet or bus stop timetable	• Information from experience or the timetable booklet
• High levels of knowledge on STOPWATCH	• Average levels of knowledge on STOPWATCH	• Low levels of knowledge on STOPWATCH
• Ignored the display or looked at it once	• Did not see the display, but the students looked at the display several times	• Did not see the display
• Mainly existing users of the route	• Some new users (students) of the route	• Existing users of the route
• Neutral view of reliability but a positive view on STOPWATCH from the full time employed	• Most positive view of STOPWATCH for students	• Generally negative view of changes in service reliability and STOP-WATCH effect
• Car available to full time, employed both as driver and passenger	• Car available to business user as passenger, but not to others	• Limited car availability either as a driver or as a passenger
• High levels of car use	• Intermediate levels of car use	• Low levels of car use

Table 7.5. Sample breakdown by public transport use.

Frequent bus users	Infrequent bus users
• Shorter waiting time	• Longer waiting time
• Catch a particular bus	• Catch the first bus to turn up
• Higher frequency of bus services	• Lower frequency of bus services
• Work and school purposes	• Visiting friends, shopping and leisure trips
• Full/part time employment	• Housewives and unemployed
• Lower level of car ownership, availability and use	• Higher level of car ownership, availability and use
• Greater use of experience	• Greater use of electronic display
• High knowledge of STOPWATCH	• Low knowledge of STOPWATCH
• Low importance of electronic display	• High importance of electronic display
• Limited increase in bus use resulting from STOPWATCH	• No increase in travel resulting from STOPWATCH
• More likely to find alternatives if waiting time is high	• More likely to remain at stop even if waiting time is high
• More likely to walk to destination or another stop	• More likely to go to local shop/bank or another stop

Those who never used the car were the elderly/retired and did not travel as much. They were most tolerant of the bus service as their perception of the accuracy of the electronic message was low, yet they had seen some improvement in bus reliability and they were likely to leave the bus stop. Those who made some use of the car have perceived little change in the reliability of the bus service or in the accuracy of the information from the display. They were not likely to leave the stop.

Table 7.6. Sample breakdown by trip purpose.

Work and school	Shopping	Visiting friends and personal business
• Car availability high for workers and low for school trip makers • Knew about STOPWATCH • Information from experience or from the electronic display • Some increase in trip making resulting from STOPWATCH • Increase in reliability of service since STOPWATCH • Average accuracy of STOPWATCH messages • School trip makers likely to leave the bus stop	• Car available as a passenger • Average knowledge about STOPWATCH • Information from timetable and the electronic display • Increase in trip making resulting from STOPWATCH • Increase in reliability of service since STOPWATCH • Above average accuracy of STOPWATCH messages • Average response with some leaving the stop	• Car available as a driver and as a passenger for personal business, but car not available for visiting friends • Little knowledge about STOPWATCH • Information from timetable booklet • Some decrease in trip making resulting from STOPWATCH • Some increase in reliability (visiting friends) • High proportion of don't knows on accuracy of STOPWATCH messages • All likely to remain at the bus stop

Overall, then the new electronic displays had made some impact on the different types of bus users. Particular social groups were more likely to respond positively than others and different users by their travel profiles were also likely to respond differently. The positive responders to STOPWATCH are young, in full time or part time employment, make frequent use of the bus (but not necessarily using the electronic displays), for education or work purposes and make some use of the car. The negative responders to STOPWATCH are the elderly and retired, who make some use of the bus for shopping and social trip purposes with little use of the car. These respondents are tolerant and conservative in their views of the bus system.

Table 7.7. Sample breakdown by car use.

Frequent car user	Some use of the car	Never use the car
• Travelling for work purpose or leisure or visiting friends • Car available for the bus trip as a driver or as a passenger	• Travelling for work or shopping purposes or personal business trips • Car available mainly as a passenger, with some availability as a driver for the bus trip made	• Travelling to shops or visiting friends • No car available for bus trips made
• Improvement in bus reliability • Not able to judge the accuracy of electronic information	• No change or less bus reliability • Accuracy of information within 1-5 minutes of time shown at the display	• Improvement in bus reliability • More than 5 minutes difference to time shown on display or not able to judge accuracy
• Not likely to leave the bus stop	• Not likely to leave the stop	• Likely to leave bus stop and walk directly to another stop or all the way

(Frequent use of car = 5 or more days a week; Some use of car = 3-4 days a week; 1-2 days a week; less than once a week; once a month or less; Never use the car)

7.3.3. The STOPWATCH Effects

The most important results from the 'After' data set relates to the STOPWATCH effect. Here, analysis has taken place for several clearly identified groups - for example, the first time users of the bus; the choice effect; and increased use of the bus. In each of these situations it can be argued that new traffic has been induced as a result of STOPWATCH. In Table 7.8, there are three types of induced traffic which relate to existing bus users - STOPWATCH switchers, frequent user effects and increased use of the bus. The total picture (Table 7.8) gives the overall induced effect of the new information system on bus patronage. Each of the groups has been allocated to one of the elements in the table. Although there is substantial overlap between the groups, Table 7.8 gives a picture of the range of different effects which could be expected from STOPWATCH.

The STOPWATCH switchers (1) is the name given to those respondents who changed their travel patterns during the journey. Because the expected delay at the bus stop was above a critical threshold, 214 (12.5%) respondents changed their travel behaviour. A range of alternative actions were made by this group of respondents. Nearly 45% switched from the bus to walk or taxi as a result of the long expected wait time for the bus (Figure 7.2).

- They are using the electronic displays in a positive way and adapting their journey so as to minimize their disutility. This may result in a change in mode (to walk or taxi), route (to another bus route) or even destination, or it may result in a diversion (to a shop prior to returning to the same or another bus stop). One of the perceived advantages of STOPWATCH is that it gives the bus user greater flexibility in responding to accurate real time information. These users had a positive view of the electronic displays and they were making full use of its potential, but they were also critical of its accuracy.

- Again it is the young and students on education trips with low levels of car availability who are likely to switch.
- This behavioural adaptation may result in a loss of patronage, but it will also mean higher levels of satisfaction and a greater feeling of control over the timing of trips. This in turn may result in more bus travel in the longer term.
- This group had a high knowledge of STOPWATCH and placed a great deal of importance on the quality of information.
- The expansion of this group of regular bus users could be achieved through careful marketing of the STOPWATCH system.

Table 7.8. New bus use induced by STOPWATCH.

'Before'	'After'	
Bus users	Bus users still using the bus - those who changed timing, route, mode or destination in course of journey - *STOPWATCH switchers(1)*	Generated traffic existing bus users making more trips - *frequent user effect (4)* and *increased use of the bus (5)*
Car users	New bus users - *choice effect (2)*	
Other modes (cycle, walk)	New bus users - *choice effect (2)*	
Not travelling	New traffic - *first time users of the bus (3)*	

Note: In the survey there may be overlap between these groups.

The choice effect (2) involved those who had a car available as a driver or as a passenger (18% of the sample), but chose to travel by bus.

- These had increased their travel as a result of STOPWATCH. On the former group, there seemed to be a limited impact from STOPWATCH, except that those with a car available as a driver had increased their use of the bus.
- Overall, it was those in car owning households that had increased their use of the bus most as a result of STOPWATCH, mainly for shopping and education trips.
- Again it was the young people who were most likely to generate more trips. The numbers were small, so not too much should be interpreted from the results.
- These respondents were mainly in the middle age groups, in full time employment, on work or personal business trips if they had a car available as a driver.
- For those who had a car available as a passenger, they were in employment or a housewife and were on shopping or personal business trips.
- The STOPWATCH effect on their trip making was limited, but they felt that the reliability of the bus service had increased.

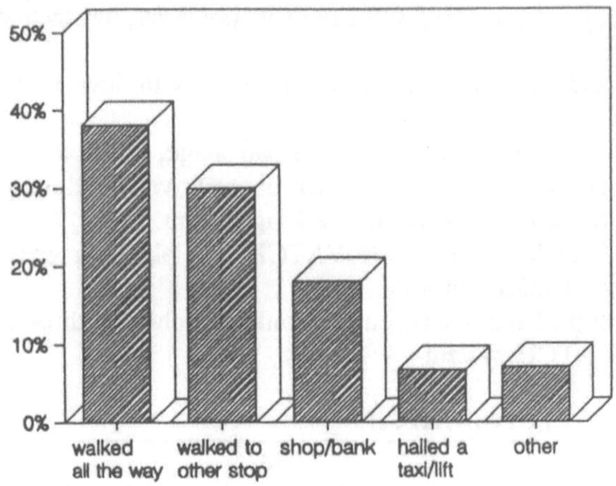

Fig. 7.2. Action on leaving bus stop.

There were two definitions of *first time users of the bus (3)* - those who never used this route to the same destination from this stop (115 respondents) and those who did not use this bus route before the installation of electronic displays (406 respondents) - new users.

- These people made more use of printed timetables and the electronic displays than the rest of the sample.
- They were mainly male in the young age categories (17-24 and the 25-44 age categories).
- These new users were positive about STOPWATCH, even though they were not the most frequent bus users.
- This mainly young cohort were students with high levels of car availability.

This is a market for bus users which can be further developed through the provision of a high quality service. Although this group had lower levels of knowledge of STOPWATCH, they had a very positive view overall to the electronic displays. Even with the wider definition of new users, there was a similar profile, but with more females and a greater student presence.

Frequent bus users (4) (40% of sample who used the bus 4 or more days a week) are less enthusiastic about STOPWATCH and they relied on their own experience for service information.

- They were likely to be women and those in employment.
- These users were dependent upon the bus with a high level of knowledge of STOPWATCH.
- They relied on their own experience for information rather than the electronic displays.
- The most important attribute of the bus service in their view was the frequency and reliability of the service.
- This group of travellers are most critical of the STOPWATCH system, particularly

on its accuracy.

- This is the most important group of bus users as they are mainly captive to the bus and use it for work and education trips.

If STOPWATCH can get through to this group of users so that they are given an accurate and reliable service, then there will be a great potential for another group of satisfied travellers to make a substantial increase in bus travel.

Increased use of the bus (5) from existing users - the generation effect. Despite the small numbers, this group was:

- Very positive towards the information system and had a high knowledge level.
- This group of bus users came from non car owning households and never made use of the car. So it was unlikely that these trips came from other modes, but were newly generated trips.
- They were likely to be male and under 34 years of age, with a high level of knowledge about STOPWATCH.

As noted above, there is considerable overlap between the categories of new bus use (Table 7.8) as some disaggregations are based on user groups, whilst others relate to trip generation effects or behavioural responses. The intention is not to produce a clear categorization, but to give a flavour of the range of responses resulting from telematics information.

As part of the STOPWATCH demonstration project, two types of display have been installed, the Light Emitting Diode (LED) displays (red on black, large displays installed above head-height in the bus shelters) and the Liquid Crystal Displays (LCD - black on grey, small displays installed at eye level).

- There is a strong preference among all respondents for the LED over the LCD when compared with printed timetable information, whether by position, ease of understanding, ease of reading or accuracy of information.
- There was however, no significant difference between the two types of display by accuracy of information or frequency of looking at the display boards (Figure 7.3).

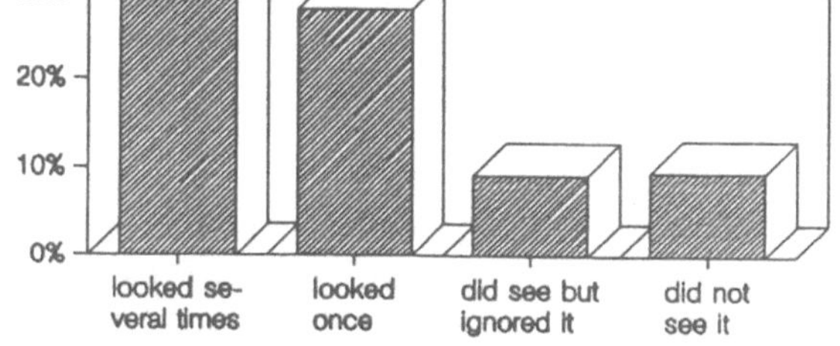

Fig. 7.3. Frequency of looking at electronic information.

It does seem that STOPWATCH has had an impact on the use of the bus with new users, some new trips being generated and some new patterns of behaviour emerging. It is difficult to isolate individual effects from the 'After' survey, so we now turn to the comparison between the 'Before' and the 'After' surveys to establish whether these patterns can be supported from the two data sets.

The difference between statements of behavioural intention and actual behaviour is well known, and the results from STOPWATCH are not unexpected (see Table 7.9). Behavioural intention typically overestimates actual demand by 3-4 times, and in the STOPWATCH corridor the figure is about 4.2 times, or an increase in trip-making of 4.25%. This difference may change in the longer term to give an increase in travel of about 5%.

Table 7.9. The STOPWATCH effect.

STOPWATCH effect	'Before' intention	'After' actual
increase	17.9%	3.7% 3.2%
decrease	0.4%	0.6% 0.7%
no effect	71.5%	94.4% 93.4%
do not know	10.8%	1.2% 2.7%

Note: The two figures in the 'After' column refer to the peak (before 18.00 on weekdays) and off peak (evenings and weekends). The overall figure for the increase in travel is 4.25% or 55 out of 1293 responses.

7.3.4 'Before' and 'After' Analysis

The disaggregate analysis and the detailed investigation of the STOPWATCH effects have presented many opportunities for analysis as there is a tremendous variability between the 'Before' and 'After' data for a range of variables disaggregated by age, gender, car availability and employment status. Sample sizes are sufficient for comparison in all situations and it has not been necessary to aggregate the sub-samples. Particular difficulties are apparent in the interpretation of the data as many of the survey responses required scaling and preferences. This may explain the variability in the nature (positive or negative) and scale of change, and the complexity of the pattern of responses. Given those limitations, a summary of the results with the data weighted by age and employment are now presented.

Aggregate analysis of the 'Before' and 'After' surveys

- *Journey Profile* shows very little difference between the two samples in terms of frequency of travel by bus or car, but it is in the trip purpose disaggregation where most change has taken place (Figure 7.4), with changes in the proportion in all purpose categories. As explained before (Section 7.3.1), this was due to the 'Before' survey being carried out in the school holidays (August 1993) and the 'After' survey in the autumn (October 1994).
- *Expected Waiting Time* had also changed. Although the proportion of respondents in

the shortest waiting time category had increased, the overall time had decreased, particularly in the longest wait categories. The availability of real time information meant that 77% of passengers in the 'After' survey waited less than 10 minutes (67% in the 'Before' survey). The perceived frequency of bus services had improved.

- *Knowledge of STOPWATCH* rose substantially over the survey period from 10% to 59%. The fact that 10% of respondents knew about STOPWATCH some 4-5 months before its introduction is a comment on the advanced publicity given to the system. Equally, the 59% figure for the knowledge, 10 months the introduction of STOPWATCH, is perhaps lower than might be expected. But it may relate to the number of new bus users and ignorance of the actual technical term.
- There were also significant differences in the *STOPWATCH effect*, as far less respondents had actually made an increase use of the system than stated in the 'Before' survey. Over 94% claimed that the STOPWATCH system had not had any effect on the numbers of trips made, as compared with 72% in the 'Before' survey (Table 7.9).

Disaggregate analysis of the 'Before' and 'After' surveys

- *Trip purpose* - tremendous change has taken place in the trip purposes, particularly in the younger age groups, and for shopping, leisure and visiting friends activities. The young made less work and visit trips, but more school, shopping and leisure trips. The 25-34 year age group made more work trips, but less school and shopping trips. The 35-44 year age group made less shopping, but more visit trips, whilst the 45-59 year age group made more shopping trips but fewer leisure trips. These variations may have resulted from the two different months in which the surveys were carried out, but it may also reflect the finding that there is considerable variation in trip patterns over time. Stability in trip making is often an assumption made in survey analysis. The Southampton data seems to question that stability.

- *Expected waiting time* - this shows an improvement across all age groups, except the elderly. For the elderly, there had been an increase in expected waiting time, particularly in the shortest time category. Students expected waiting time had declined substantially between the two surveys.
- *Frequency of bus service* - there was a substantial increase in perceived frequency of bus service, particularly by those in full or part time employment. The unemployed thought that service quality had declined. Improvement was also seen by those with a car available, either as a passenger or as a driver. In the 'After' survey, 28.2% of those with a car available stated that service frequency was under 10 minutes, whilst the corresponding figure in the 'Before' survey was only 6.3%.
- *Source of information* - employment status was again significant with an overall decline in the use of timetable information. There is some substitution effect, but also a stimulation effect to use more information. Use of the electronic display is close to expected for all groups, except the students who have a higher than expected use level.
- *STOPWATCH effect* - there was some overestimation of the STOPWATCH effect by housewives and students and an underestimation by the unemployed and retired.
- *Attributes of the bus service* - the disaggregate analysis has been restricted to the

three significant attributes found at the aggregate level. The importance of each attribute (printed timetable, electronic displays and waiting facilities) had declined over time. For the young, the importance of the printed timetable had increased, as had the importance of the electronic displays. The elderly increased their responses in the low importance category for the electronic displays.

Before

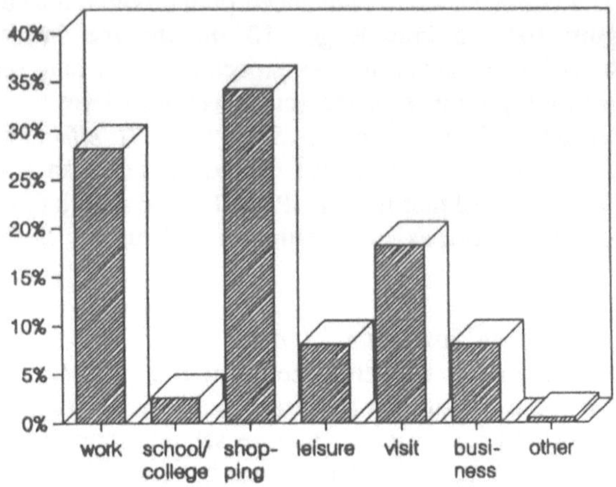

After

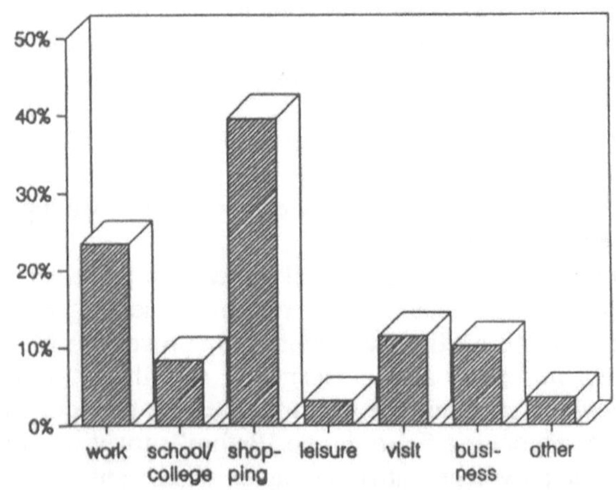

Fig. 7.4. Trip purpose 'Before' and 'After'.

7.3.5 Summary of Results from Southamptom STOPWATCH Surveys

Apart from establishing that on some variables there were significant differences between the two samples, some interesting results are already apparent. The main analysis needs to be carried out on a weighted 'Before' data set to ensure that the socio-economic structure of the two samples is similar. If the socio-economic structure is controlled, then the changes in behaviour, knowledge and perceptions can be attributed to the STOPWATCH system, rather than to the structure of the sample. The weighting procedure was carried out by age and employment status as these were the two variables which were significantly different.

From this preliminary analysis it was clear that:

- Trip purpose differed significantly between the two samples. The explanation here was the timing of the surveys with the 'Before' data being collected in August when the number of education trips is reduced, whilst those for leisure and visiting friends and relatives are increased. Conversely, the 'After' data was collected in October when education trips are increased, with other purposes (particularly leisure) being reduced. A common survey month could have controlled for this variation, but the range of trip purposes adds richness to the data analysis. Even with the adjustments for socio-economic profile in the weighted data, trip purpose variations are still noticeable between the two samples.
- The STOPWATCH effect was also clear. Bus users felt that a frequent and reliable service was the most important attribute of the service and the new source of information met part of that requirement.
- Knowledge of STOPWATCH was only admitted by 60%, even after the system had been in operation for 10 months, but it had already had a significant effect. Responses were positive to the STOPWATCH system, and it seemed to replace the use of timetable information at the bus stop and published timetables.
- There was a strong preference for the larger LED displays than to the smaller LCD displays.
- Having better information may make bus users more critical in general of the quality of public transport information. STOPWATCH has raised expectations.
- There were also a substantial number of new users of the bus, and the real time information had allowed 12.5% of the sample to change their destination, mode, route or timing of their current trip (Figure 7.2).

All of these issues have been taken up in the second part of the analysis which presents a disaggregate analysis of the 'After' surveys. The main results from the data are:

- There are differences by gender which reflect wider factors than just the impact of STOPWATCH. There were more men who were new users of the bus corridor (25%) than women (21%).
- However, as with much of the analysis in this report, age is the key discriminating social variable on use of buses. The young (17-24 and 25-34) were most likely to have increased their trip making as a result of the STOPWATCH system, they were more generally satisfied with the reliability of the bus service and the accuracy of the STOPWATCH messages. They were also more likely to leave the bus stop

because of the expected long delay until the next bus arrived, and to make use of the greatest range of actions. The elderly were the opposite. They were more modest in their views and less likely to make changes in their travel behaviour.

- The employment differences mirrored the disaggregations by age. It was the retired who made least use of STOPWATCH or did not even see the signs. But even this group thought that the information was more accurate than the printed timetable. The students were most likely to leave the bus stop, whilst the other employment groups were more conservative in their actions.
- Frequent bus users were again likely to be more positive in their assessment of the services provided, as they had shorter waiting times and stated that frequencies of services were good. These users were mainly captive to the bus and were in employment using the services to go to work or school. They made most use of their own knowledge and had a high level of knowledge on STOPWATCH, but gave a low level of importance to the electronic display. These users had made modest increase in use of the buses since the introduction of STOPWATCH and were likely to find alternatives if the wait time was high.
- Infrequent bus users had a less well informed view of the service with longer wait times and an expectation of lower frequencies of service. These people also had a car available, but travelled on the bus for non work related activities. Although they had a low level of knowledge of STOPWATCH, they made a greater use of the electronic display, but this had not resulted in more use of public transport. They were likely to remain at the bus stop, even if the expected delay was substantial.

The conclusion here is that bus use is a key indicator of the actual and expected changes in behaviour resulting from the introduction of STOPWATCH. As it links in with the new user effects, it is recommended that these two variables be used as the main monitoring variables for the STOPWATCH effect. The other factors making up the journey profile, namely trip purpose and car use reinforce the conclusions drawn on bus use.

When a comparison is made between the two surveys in Southampton, travel patterns in terms of use of the car and the bus shows a remarkable consistency:

- It is only with trip purpose that significant differences occur, particularly the increases in education and shopping trips, and the decrease in work, leisure and visiting friends. The explanation is probably the timing of the two surveys and the effects that the Summer holidays have on travel purposes.
- Overall, expected waiting time for the bus has been reduced, but short wait times have increased.
- Similarly, the perceived frequency of bus services has improved, and bus users are more likely now to set out for any bus rather than one particular bus. One explanation for these changes may be the availability of real time information as this reduces uncertainty and raises awareness of how frequent the bus service actually is.

Changes in the STOPWATCH effect may in part be related to the points made above:

- The electronic displays have replaced the use of published timetables and timetables at stops.
- The use of experience or prior knowledge has increased.
- The potential impact of STOPWATCH was much greater than the actual impact.

Intention had overestimated actual change by about 4.2 times. This is higher than expected and may be reduced to about 4 times, which will result in a total increase in demand for buses of about 5%.

In the disaggregate analysis, the significant variables were investigated in detail:

- Trip purpose has changed substantially over the two surveys, and needs to be carefully monitored over a longer period. Most travel surveys are cross sectional at one point in time and claim to represent a typical day or week. However, it seems from the Southampton data that there is substantial variation between the two survey months, not in the bus trips made but in the purpose of those journeys. Longer term monitoring should cover the numbers of bus trips made, their origins and destinations and their purpose - the surveys would be carried out at different times of the year. It seems that bus travel patterns by trip purpose are not as stable as once thought.
- Expected waiting time has improved overall, but there is a substantial variability between different age groups with the young (17-24) and the elderly (>60) showing the greatest changes.
- Frequency of bus service had improved overall, particularly for the 25-34 and 45-59 age groups and for those in employment or retired with a car available.
- There has been an overall decline in the use of timetables, but this change is not uniform across all groups. In some cases there is a substitution effect of electronic displays for more traditional forms of information, but in other cases there has been a general increase in the amount of information used by all bus travellers, particularly students.
- The STOPWATCH effect shows a clear difference between the 'Before' and 'After' surveys by employment with some groups having overestimated the effects (e.g., housewives and students), whilst the unemployed and the elderly have been more cautious in their responses.

The disaggregate analysis demonstrates the power of 'Before' and 'After' survey analysis, and the possibilities for variability within and between the data sets, even with a limited breakdown of the data. A minimum of 30 cells have been used in the analysis to maintain sample sizes.

7.4 Conclusions from the Case Study

It is not often that 'Before' and 'After' survey data are collected so that analysis can explore the difference between stated intentions and actual behaviour. Two recent studies of the user response to new transport telematics applications have been presented in this chapter. The Birmingham survey is relatively limited in its scope as the sample sizes were small and as the 'After' surveys were carried out immediately after its implementation. This has not allowed the system to stabilize, nor has it permitted behavioural response to be consistent - it really only records immediate reactions to the passenger transport information systems. This reaction is favourable, but it has not yet been translated into new trips.

The Southampton surveys demonstrate a much more mature reaction as the passenger transport information systems (STOPWATCH) have been in operation for 10 months. The scale of the passenger transport information system is more extensive and a much higher profile has been given by the local authority through publicity and the integration of these new systems into its overall transport strategy.

In Southampton, behavioural surveys have been carried out on bus users before and after the installation of the real time information system. About 11% of the before sample (1538 respondents in total) were aware of the plan to introduce the system, with those who were frequent users having greater knowledge. About 16% of respondents stated that the STOPWATCH system would increase their use of the bus, but that this would be most significant among occasional and first time bus users. The most positive responses came from those who used the bus for social and shopping purposes, particularly among the young. However, when compared with the survey after the installation of the passenger information system, very different results were apparent. In the after survey were 1702 respondents and only 4% had actually increased their trip making as a result of the STOPWATCH system. There seem to have been an overestimation of the impact on the trip making patterns of bus users about three times. The difference between behavioural intention and actual behaviour is clear.

Table 7.10. Summary of the STOPWATCH effect in Southampton.

Travel characteristics	New bus users	New trips from existing bus users	STOPWATCH switchers	All sample
Trip purpose				
- work	20.7%	25.9%	21.5%	23.5%
- education	14.6%	15.7%	13.6%	8.4%
- shopping	34.1%	45.1%	40.2%	39.6%
- other	30.6%	13.4%	24.7%	28.5%
Car available	47.5%	19.7%	13.2%	17.8%
Non car owning households	52.5%	68.2%	57.9%	55.5%
Never used the car	15%	24.7%	18.5%	14.8%
Socio-economic characteristics	**New bus users**	**New trips from existing bus users**	**STOPWATCH switchers**	**All sample**
Women	56%	60.7%	63.6%	63.1%
Under 34 years	69.7%	67.7%	62.2%	50.3%
Full-time employment	28.8%	40.7%	34.1%	38.9%
Full-time student	37.2%	25.1%	40.1%	23.3%
Sample size	406	55	214	1702

Nevertheless, there were three main ways in which STOPWATCH seemed to have had an effect on travel behaviour. There are a substantial number of new users of public transport, since the introduction of STOPWATCH. These new users were young, usually travelling for education or leisure purposes, and they often had high levels of car availability. They had a very positive view towards STOPWATCH, but made a less than average amount of use of the bus service. The new users seem to form the basis of a revitalized interest in new types of public systems, and the importance of high quality real time information to this group of users. The marketing of STOPWATCH to this group is important to the success of information systems as these new users have generated more bus travel (Table 7.10).

The second STOPWATCH effect has been to encourage existing bus users to make more trips on the improved service. This 4% mentioned earlier have a high level of knowledge of STOPWATCH (73% as against the sample average of 59%), and they actually make use of the electronic display information (35% as against the sample average of 22%). All of this group's respondents used the route before the electronic displays were installed, and over 80% looked at the displays on several occasions (sample average 53%). This group of users came predominantly from non car owing households and they were never likely to use the car. They were young people on shopping or education purpose trips. It seems to be possible to increase the use of buses, even among captive users and heavy users of the existing services through marketing a better quality service with real time information systems. These increases in bus use are small, but as the overall quality improves, further trips should be generated.

The third effect is the STOPWATCH switchers. These are bus users who have left the bus stop during the last fortnight because the electronic display showed that the bus would not arrive for some time. Some 214 (12.5%) responded positively in this way and they either switched route to another bus or stop, or went to a local facility and then returned to same stop or another stop (Figure 7.2). The total mode shift effect resulting from these figures is 45%.

So, actions can involve modal changes, route changes, timing changes, activity changes or destination changes. The introduction of the STOPWATCH system has allowed a new flexibility in trip making that seems to have been well received, particularly by the young people in education. However, there is also a concern over the accuracy of the information and the most important issue for all bus users is a reliable and frequent service.

The use of technology in bus services provides a better quality service, particularly in a deregulated bus market such as that in the UK. One of the major losses caused by deregulation has been the poor quality of information about the bus services which are available. The numbers of services running in the Southampton area and their frequency has increased, but that information is not available in the bus. Good quality real time information does not seem to improve the perceived quality and reliability of the bus service. Moreover, even after only 10 months of the STOPWATCH system in Southampton, these users are making more journeys, although on a modest scale.

Annex 7A Econometric Analysis of STOPWATCH Pilot

In general, modal choice is determined by various factors, such as monetary (out of pocket) costs, travel time (including waiting time for public transport or in traffic jams), (in)convenience of the travel mode used, flexibility, safety etc. The aim of this Annex is to provide a solid economic basis for the trade-offs to be made in the modal choice, with particular emphasis on the question how provision of information on expected arrival (or departure) times of public transport (with particular emphasis on buses) impacts on the travel mode chosen by trip-makers.

Information provision can actually have two distinct effects. The first is that it gives travellers better prospects of making optimal choices among alternatives; secondly it reduces the uncertainty involved in waiting: the (psychological) costs incurred by waiting will be less when the traveller knows exactly how long he has to wait. Exact knowledge may, for example, permit the traveller to start some alternative activity without the risk of missing the bus. We will analyze both mechanisms, using an elementary model.

The basic model

Suppose a traveller T wants to make a trip from a starting point A to a well defined and single destination B. Suppose T faces two alternative modes viz. bus and another one which will for convenience be called car. Assume that T is a cost minimizer and let C_c be the cost of using the car and C_b the cost of using the bus. Throughout this paper we will assume that C_c only consists of a fixed amount FC_c, which is unaffected by the information provision that is studied here.

The costs of using the bus on the other hand are assumed to consist of three parts: a fixed cost FC_b, the cost of waiting per se, and some cost associated with the uncertainty involved in waiting. The inclusion of waiting time cost is based in an opportunity cost argument. The cost due to uncertainty is explained above, and is basically also a measure of opportunity cost. This may seem to be a sort double counting but it is not. The reason to include uncertainty as a cost measure is that uncertainty factually prohibits an alternative activity during waiting. Nevertheless, the utility associated with such an alternative is by definition less than the utility that should be derived by the first preferred use of the time involved, which is travelling. Thus, waiting involves cost as the first best alternative is unattainable, and the possibility to choose a second best alternative during waiting will reduce this cost, but not fully compensate for it.

Accordingly, the cost of taking the bus for a given trip can be formulated as

$$C_b = FC_b + \alpha \, WT + \beta \, \sigma^2 \qquad (A7.1)$$

where waiting time WT is multiplied by a factor α ($\alpha > 0$) which is a measure of the time value, and the variance of the waiting time σ^2 is a cost measure of the uncertainty, weighted by a factor β ($\beta > 0$). Traveller T will compare these costs with FC_c the generalized cost of the alternative.

The effect of uncertainty reduction

First we will concentrate on uncertainty reduction, neglecting the waiting time WT.

As T is a cost minimizer, for some given trip he will choose the bus when

$$FC_b' + \beta\,\sigma^2 < FC_c$$

or

$$FC_c - FC_b > \beta\,\sigma^2 \tag{A7.2}$$

Next, assume that T considers a large number of trips during a given time period and that the term FC_c - FC_b involved in all these trips follows some probability distribution over an appropriate interval. This distribution has a mean υ_T, which depends on the characteristics of T, i.e. each traveller will have another specific distribution.

Suppose for example that this distribution is normal with mean m. This is illustrated in Figure 7A.1a, where m > 0. Referring to condition (A7.2) it is obvious that T will choose the bus for all cases where the difference in fixed cost is right from the point $\beta\sigma^2$ on the x-axis, which will be called the modal-split point (m-s point).

The effect of reducing the uncertainty is a shift of the m-s point to the left (see Figure 7A.1b), and when reduction is interpreted as removing, the new m-s point coincides with the origin. The share of trips for which a change of mode occurs is given by the shaded area. Consequently, the impact of the reduction in uncertainty is dependent on three factors: first, the magnitude of the original uncertainty σ^2; second, the magnitude of the parameter β - these two factors together determine the size of the shift of the m-s point along the x-axis; and third the location (and shape) of the distribution function.

Clearly, it is difficult to tell how this distribution will look like and hence inferences can only be made on basis of rather far reaching assumptions (like normality). However, the notion that the effect also depends on an unknown distribution function is quite important. The most noteworthy consequence is that it shows that the largest effect of a policy measure providing better information is not necessarily to be expected to occur with those travellers that do not often travel by bus. An ad hoc reasoning could easily lead to this expectation as less frequent bus travellers have a large potential of using the bus. The model presented here suggests that this result would indeed follow when the m-s point is far to the right of the origin, i.e. when β or σ^2 is large, which is a specific case (illustrated by Figure 7A.1b). However, when the modal split point is close to the origin, i.e. $\beta\sigma^2$ is relatively small, the location of the distribution function may be located such that the respondent is still a less frequent bus user, in particular when the mean of this distribution is relatively far to the left. The effect of uncertainty reduction will then still be small (see Figure 7A.1c). Consequently, both the location of the m-s point and the location of the distribution function are important in determining the effect of uncertainty reduction.

The second effect that can result from information provision is an improvement in choosing among types of public transport given the choice of a mode. In particular, we will assume here that the traveller has a multiple choice at the bus stop and that the available information informs him about the arrival times of two types of buses. Our goal is to analyze how the improved optimality of the choice between these types can affect the attractivity of the mode in general. Concentrating on the variable WT in model (A7.1) we further neglect the uncertainty term $\beta\sigma^2$.

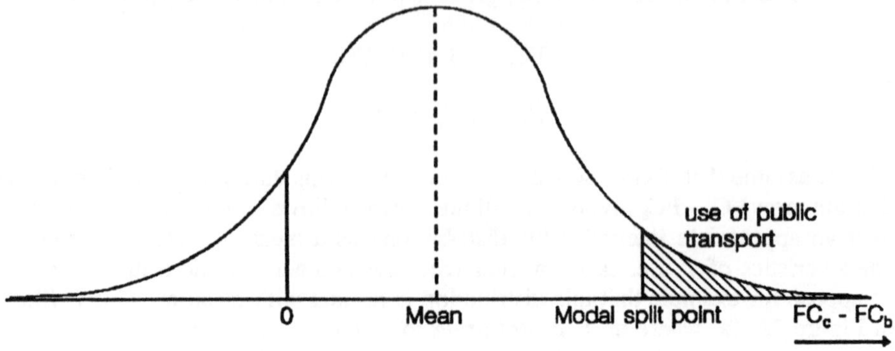

Fig. 7A.1a. The modal split with uncertainty.

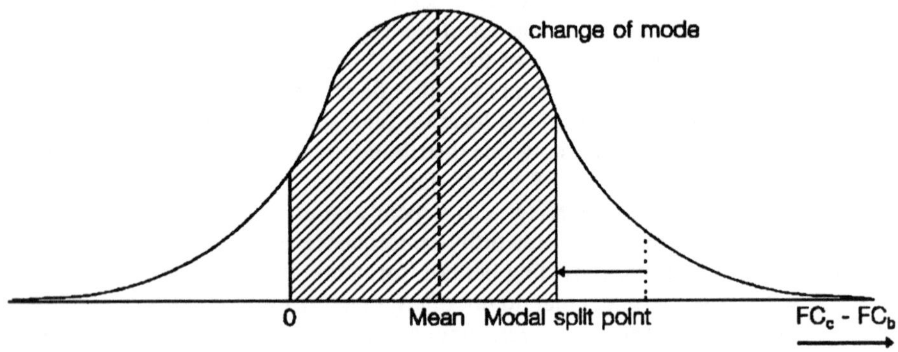

Fig. 7A.1b. Change in modal split by uncertainty reduction.

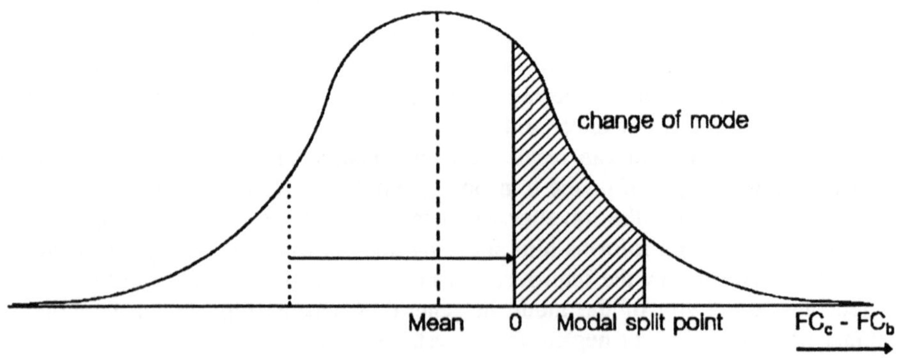

Fig. 7A.1c. Influence of location of distribution.

The effect of improved optimality in type choice

Thus we assume that T can choose among 2 alternative types when taking a bus - B1 and B2, respectively - in addition to the car alternative. The cost functions for the buses are now given by

$$C_{Bi} = FC_{Bi} + \alpha\, WT_i i = 1,2 \tag{A7.3}$$

where WT_i represents the waiting time for bus i.

Clearly, the choice of a particular bus is conditional on the decision to choose the bus anyway. In other words, T follows a two-step decision process: first he chooses between car and bus, and next, after having chosen the bus he chooses (in practice at the bus stop) among the two available types of buses. Hence, the problem T has to solve can be formulated as:

$$\text{Min } \{ C_c, C_b \} \tag{A7.3}$$

where C_b is the cost that results when choosing among the types of buses, given that the bus alternative is chosen. Clearly, T will actually assess expected costs, since part of the costs are uncertain at the moment he has to make a decision.

It is evident that information provision will only have a positive effect on bus use, if it reduces the (overall) expected cost C_b of using the bus. Therefore, the decision at the bus stop has to be further analyzed.

Thus, consider now the situation that T is at the bus stop, and is optimizing his (type of) bus choice problem. The costs T will incur are given by (A7.2), and without loss of generality it is assumed that $FC_1 < FC_2$, i.e. option 2 is the more expensive alternative. Instead of indices 1 and 2, for reasons of clarity, we will use the indices c for cheap - which should not be confused with the index c we used for car before - and e for expensive, thus, $FC_c < FC_e$ by definition.

Let t_c and t_e be the waiting times for the cheap and expensive bus, respectively. Clearly, T will choose the cheap bus if $t_c < t_e$ - i.e. the cheap bus arrives first - since then both cost elements (fixed cost and waiting time) are smaller for this alternative. Moreover, in certain cases T may choose the cheap bus even if the expensive one arrives first, and hence an extra waiting time cost is incurred. This will in particular be so when the expected remaining waiting time for the cheap bus is sufficiently small so that the extra cost of taking the expensive bus do not outweigh the cost of a longer waiting time.

To proceed, we suppose that the waiting times for both bus types are random variables with density functions $f_c(t_c)$ and $f_e(t_e)$, respectively, both well defined on the interval $[0, \infty)$. Consider the case that the expensive bus arrives first at t_e. Taking the expensive bus would yield the following costs:

$$FC_e + \alpha\, t_e \tag{A7.4}$$

However, T can also wait for the cheap bus to arrive to enjoy lower fixed costs, yet incurring more waiting time costs which are moreover uncertain. Indeed, the expected (total) costs of waiting for the cheap bus are:

$$FC_c + \alpha \int_{t_e}^{\infty} t_c f_c(t_c \mid t_c > t_e) dt_c \tag{A7.5}$$

Thus, T will take the expensive bus if it arrives first *and* (A7.4) is less than (A7.5). Otherwise, T will take the cheap one. By putting (A7.4) equal to (A7.5) and solving for t_e, a break-even point, say t_e^*, is found at which T is indifferent between the two buses (a unique solution is assumed). Obviously, for $t_e < t_e^*$, the expensive bus will be chosen as the expected additional waiting time for the cheap bus is longer than is warranted by the reduction in fixed costs; for $t_e > t_e^*$ it pays to wait for the cheap bus. This is illustrated in Figure 7A.2 which shows the arrival times, or waiting times for the expensive bus on the vertical axis and for the cheap bus on the horizontal axis - the origin represents the point that T arrives at the bus stop. On the t_e-axis the point t_e^* is given. The shaded area indicates the arrival times of both buses, such that $t_e < t_c$ (under the 45° line) and $t_e < t_e^*$ (under the indicated point). Hence the shaded area indicates the combination of arrival times for which the expensive bus is chosen. Obviously, for some of the combinations the chosen bus is not ex post optimal, and information provision can help in preventing T to make the wrong bus choice in these cases. This is now analyzed below.

Suppose that the expensive bus arrives first at t_e, and that T gets information about the (expected) arrival time of the cheap bus. Hence, T will now compare (A7.4) with a new expression for the expected cost of taking the cheap bus:

$$FC_c + \alpha\, t_c \tag{A7.6}$$

In this case t_c is not a random variable any more, but the arrival time revealed by the information. Comparing (A7.4) with (A7.6) gives a condition for the arrival times, so that the expensive bus is chosen. This condition reads as:

$$\alpha\, t_e + FC_e < \alpha\, t_c + FC_c$$

or

$$t_c > t_e + (FC_e - FC_c) / \alpha \tag{A7.7}$$

Notice that this latter condition implies $t_c > t_e$. Figure 7A.3 shows the combination of arrival times for which the expensive bus is chosen, in a way similar to Figure 7A.2. In order to compare Figures 7A.2 and 7A.3, as the next step Figure 7A.4 is constructed. This figure shows the effect of the information provision and can be used to demonstrate that information unequivocally has the effect of decreasing the expected cost of using the bus.

In Figure 7A.4 the (t_e,t_c) space is divided in four areas. The unshaded area I represents all points for which in both regimes (with or without information provision) the cheap bus is chosen. Similarly, the horizontally shaded area II contains all points where in both regimes the expensive alternative is chosen. The diagonally shaded area III, indicates that the expensive bus is chosen when there is no information provision at the bus stop, whereas the cheap bus is chosen when information provision is in effect. For these combinations it is clear that taking the expensive bus is suboptimal, since the extra waiting time does not exceed the difference in fixed cost. Finally, the diagonally shaded area IV shows the reverse case. For these combinations the cheap bus is chosen without information provision, whereas the expensive alternative is chosen in the opposite case. Again it is easy to observe that without information provision a suboptimal choice is made.

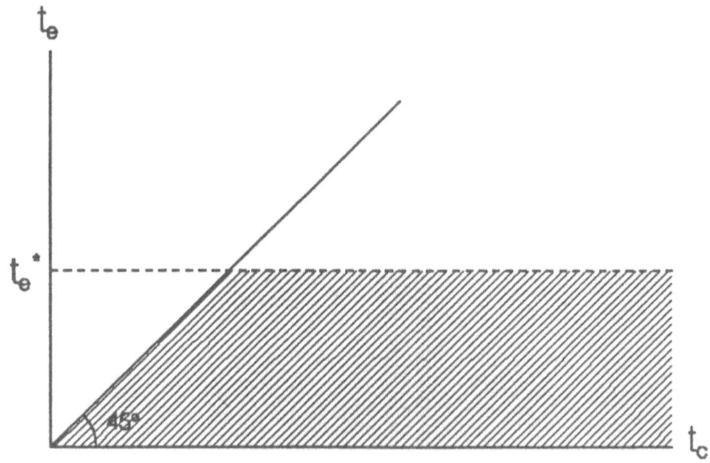

Fig. 7A.2. Bus choice without STOPWATCH.

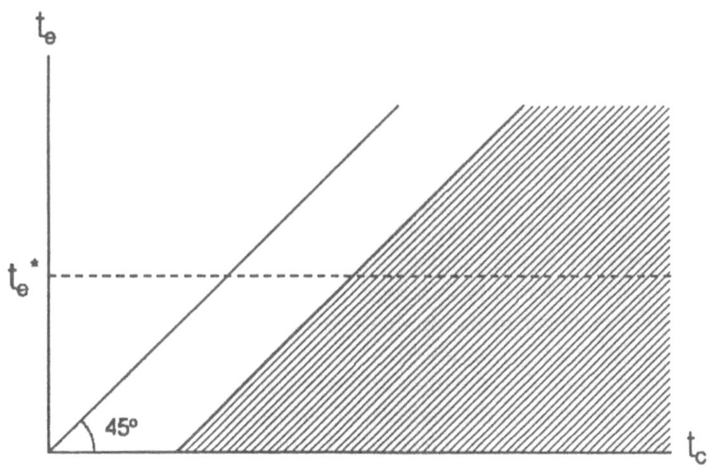

Fig. 7A.3. Bus choice with STOPWATCH.

Therefore, it can be concluded that the effect of information provision is a better choice of bus type - i.e. with less expected cost - when the combination of arrival times falls in either one of the two diagonally shaded areas of Figure 7A.4. Hence, with information, the traveller is always at least as well off compared to the case without information provision, and in some cases he is better off. An interesting question is now, how this effect depends on other features of the problem, in particular the α

coefficient representing the value of time. In Ouwersloot et al. (1995a) it is shown on the basis of this model that the largest effects may be expected for intermediate α's, while for very small and very large values of α the effect will be zero. In particular for large α this is surprising as intuition suggests that large effects of reduction in waiting times are likely to be found for those with high time preferences, which is essentially the interpretation of α.

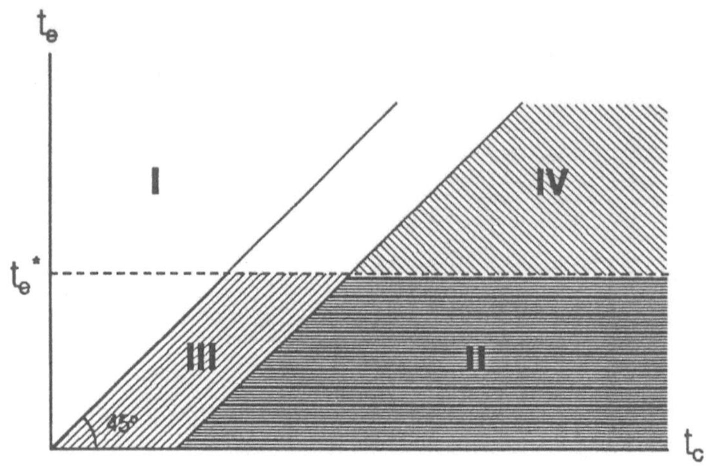

Fig. 7A.4. The STOPWATCH effect.

Empirical results

The analysis above suggests four relationships that can be investigated with the data from the STOPWATCH project. From the analysis of uncertainty reduction, it was predicted that the effect of STOPWATCH will be larger when:

i) the uncertainty of arrival times is greater;
ii) the cost attached to this uncertainty is larger;
iii) the modal split point is closer to the mode of the distribution which describes the difference in fixed cost of the trips involved.

The analysis of the improved choice situation at the bus stop, led to the prediction of

iv) a parabolic relationship between the increase of bus use due to STOPWATCH and the value of time, with a maximum at intermediate values.

These four relationships were investigated by means of applying a simple logit model for both the 'Before' and the 'After' data. A full account of model choice and related issues is given in Ouwersloot et al. (1995a); here we only report the results in Table 7A.1.

Relationship i) is not strongly supported by the estimations. It is captured by two variables 'frequency' and 'time table knowledge' both of which are expected to have a positive sign. In the 'Before' estimations this positive relationship is found, but only partly significant. However, in the 'After' estimations, for both a significant *negative* parameter is found. On the other hand, the 'Accuracy' of the STOPWATCH technology, clearly only measurable for the 'After' data, gives strong additional support for relationship i). The attitude towards uncertainty (rephrasing relationship ii) is analyzed by the variable 'IMP_TF' for which again a positive estimate was expected. The result may be called inconclusive. The estimate is positive but by no means significant. Finally the variable 'Sharebus' is included to study the non-linear relationship iii). For the quadratic term in particular a negative estimate is expected, which is indeed found for the 'Before' data, whereas the 'After' estimate is positive but not significant. Hence some support for this relationship is present.

Table 7A.1. STOPWATCH effect for full model.

Dependent Variable: Stopwatch Effect = 1 if respondent reports an expected increase in bus use due to the introduction of Stopwatch.

	Before		After	
Variable	Parameter	St. Error	Parameter	St. Error
Constant	-2.97	.83**	1.67	3.21
Male	-.29	.15*	-.11	.30
Age 25-34	-.22	.21	-.13	.40
35-44	-.60	.24**	-.57	.45
45-59	-.31	.21	-1.41	.62**
60+	-1.21	.22**	-.70	.43*
Frequency	.0009	.0030	-.046	.025*
TimeTable Kn.	.24	.086**	-5.15	2.51**
IMP_TF	1.71	1.38	2.68	2.29
Sharebus	-.016	.050	.072	.10
Sharebus^2	-.011	.016	.011	.038
Accuracy			-.67	.16**
EWT	.13	.09	.44	.20**
EWT^2	-.0055	.0048	-.025	.011**
Number of observations		1494		1250
Log-Likelihood		-626.43		-196.45
Log-Lik UNC_RED (a)		-628.20		-199.18
Log-Lik IMP_OPT (b)		-634.23		-212.49
LRT (a) (2 df)		3.54		5.48*
LRT (b) (5/6 df)		15.60**		32.08**

* means significance at 10% level
** idem at 5% level

The value of time is supposed to be related to the 'Expected Waiting Time', and thus this EWT variable is also supposed to show a parabolic relationship to increase in bus use, with a maximum for intermediate values (which should be revealed by a negative sign). This is indeed found in both estimations, which are in the 'After' case moreover strongly significant.

Finally the estimation statistics suggest that most of the increase is explained by factors related to the improved possibilities to make an improved choice at the bus stop, whereas less is explained by the possibilities to use waiting time more efficiently. Thus, although the 'improved optimality' at first sight is a highly theoretical effect, it nevertheless deserves a serious approach.

8 Telematics Use by Road Freight Operators: A Dutch Case Study

8.1 Introduction

Road freight operators show a considerable variety of size classes and forms of operation in Europe with different organizational and operational problems to overcome. Small hauliers constitute the vast majority of the sector, but when considering the number of vehicles operated, it is clear that the large hauliers play a dominant role. A great variety of market segments can be distinguished within road freight operating companies, based on different criteria (e.g. product type transported, operation type etc.). Significant changes are expected in the freight sector in the EU in the coming years (Bollo 1992). Some important developments which need to be mentioned in this respect are the completion of the European internal market, the related deregulation of the European and national freight markets, and the changing requirements of major users of freight services. These developments will likely have advantageous impacts such as the simplification of trading procedures, reduction in trading and transport costs, removal of cabotage restrictions and thus a better access to markets. In this light we may mention the evolution of 'mega-companies' which will offer full logistics services with also other transport modes (Cooper 1992). In fact, these integrated carriers are expected to be the architects of future transport by investing large amounts in informatics, while sub-contractors will be mainly concerned with the actual trucking.

Considering the above developments and combining this with the increasing congestion levels on Europe's roads, it is clear that in addition to the collective transport telematics users like road managers, road freight operators are a very important potential actor at the demand side of the transport telecommunications market. We may expect that in this sector telematics innovations may be adopted earlier than in the case of private travellers because the economic interest is greater and transport firms may have greater opportunities to invest in the necessary equipment because of their larger financial possibilities. In contrast to the public sector (road managers), the adoption of new technological innovations is in the case of road freight operators dependent on clear economic criteria rather than on issues of public or social interest. Investment costs of these technologies will simply be compared with their expected advantageous impact on operating costs.

The requirements of road freight operators regarding the use of information technologies are varied. A list of such needs is shown in Table 8.1. These requirements can be met by a large range of telematics applications currently available or still in development (OECD 1992). These systems include computer generation systems of goods (e.g. by bar coding and radio tags), electronic data interchange (EDI) systems, automatic vehicle location (AVL) systems, mobile communication systems and navigation systems. Clearly, the nature of EDI is quite distinct from the other mentioned telematics applications. Yet there are two reasons to treat EDI technology on a comparable basis: first, EDI is a true telematics application, i.e. an offspring from the marriage between telecommunications and computer technology; and second, like the others, EDI is aimed at reducing average round-trip time, economizing on operating costs.

We will in the remainder of this chapter focus on EDI, AVL, mobile communication and autonomous navigation systems.

Table 8.1. Transport telematics requirements of road freight operators.

Freight & logistics management	
Planning:	Long term-transport planning. Transport demand/resources. Transport offer calculation. Transport order.
Execution:	Notice of dispatch. Transport documents. Customs clearance.
Controlling:	Cargo tracking. Shipment status. Transport status. Delivery information.
Administration:	Cost and performance follow-up. Invoices and payment. Post-analysis of logistic performances.
Fleet management	
Planning:	Fleet monitoring. Tactical route planning.
Execution:	Operational route planning. Load plan. Preparation plus transfer of documents.
Controlling:	Route guidance and traffic information. Vehicle/cargo tracking. Fleet monitoring.
Administration:	Cost and performance follow-up for vehicle fleets.
Vehicle management	
Planning:	Vehicle preparation. Traffic/weather information. Trip preparation. Cargo/storage control. Documentation control.
Execution:	On-board vehicle/cargo monitoring. Trip route selection. Automatic debiting. Consignment delivery/pick-up. General communication. Emergency calls.
Controlling:	Recording of vehicle performance data, cargo data, trip data. General information.

The main applications identified generate a variety of benefits for the road freight operator. Next we will discuss concisely these benefits. This will give some clues as to which firms may be expected to be the (early) adopters of these technologies. In Table 8.2 for each application the main telematics functions and the potential benefits to the operator are listed.

EDI is the transmittance of freight accompanying documents by electronic means, guaranteeing that these documents arrive at the locations where they are needed (delivery points, but also national borders) before the truck arrives. Consequently, the control at these points can be prepared in advance, and can therefore be carried out much more efficiently, thus reducing waiting time for the drivers. Another advantage of EDI, but also a precondition for its use, is that standard formats are used, reducing the chances of errors and interpretation difficulties to an absolute minimum.

Table 8.2. Main functions and benefits of telematics systems in freight and fleet management.

Innovation	Functions	Benefits
Electronic data interchange	Transmit business data and provide electronic business documents - purchase order - bill of loading - packing slip - invoice - electronic funds transfer	- efficient electronic ordering and billing - pick-up and delivery verification
Automatic tracking and tracing	Transmit information on vehicles and products - identification - size and weight - origin and destination - vehicle type or class	- traffic counting and vehicle classification - compliance with regulatory requirements - automatic toll collection - efficient sales and inventory - tracking of shipments and status checking
Automatic vehicle location and two-way communication	Exchange messages between dispatcher and driver - trip and shipment information - location (of vehicle, destination, etc.) including location of maintenance and repair shops	- manage logistics while in transit - arrange repairs for breakdowns in transit - quick response to emergencies
Autonomous navigation	Provide driver information - highway and traffic conditions - location (of vehicle, destination, etc.) - alternate routes - automatic vehicle spacing - blind spot warning	- identification of most direct or least time route - avoidance of incidents and congestion - avoidance of road hazards particularly during bad weather

While the economic benefits of EDI for firms are quite clear, this is less so for the other telematics applications. Both AVL and mobile communication systems can improve to a certain level the efficiency of transport operations. First, these systems can eliminate costs and delays associated with frequent truck driver 'check calls' to their dispatchers. These costs are especially high on irregular routes across long distances. Each of these calls requires the driver to find parking place for the vehicle, wait in a queue for an available payphone, etc. Each one of these check calls can easily consume 45 minutes (OECD 1992). The use of automatic vehicle location and long-distance, two-way communication systems eliminates the need for virtually all check calls. The dispatcher knows (usually with an accuracy of about 30 km) the location of the vehicle, so that the driver does not have to call in to report progress. In addition, logistic changes like locations of new loads, changes in delivery times and destinations etc. can be communicated to the drivers on their way. Second, these systems can reduce load-waiting times and empty miles of travel by allowing dispatchers to better match available loads with appropriate equipment. Third, the use of these systems may strengthen the competitive position of the freight operator. As more and more firms demand real time deliveries to meet new customer demands better, the exact knowledge of the status of shipments becomes a more crucial part of the transport operator's services.

Some differences exist however between the implications of AVL and mobile communication for the road operator. AVL systems are at first sight in the home-office's interest in particular. It is however difficult to think of any benefits to the firm when used without having a means to communicate with the driver as well. So it seems logical that AVL will almost exclusively be used in combination with interactive communication media. An exception may be those firms with drivers on routes who are making frequent stops where they can be called, presumably in a densely populated area.

Mobile communication systems offer the possibility for the home-office and the driver to be in permanent contact. This implies in the first place that all relevant information can instantaneously be exchanged, so that both driver and office can be provided with complete information. This full information is of course only useful if decisions - or adjustments therein - are likely to be made during the trip. There are several relevant situations in which this may be the case:

- the driver has to pass some highly congested routes, for which alternatives are available;
- the complete route (in particular the return trip) is not fixed beforehand;
- there is a high chance that situations arise in which the driver needs office information or guidance, e.g. conflicting situations with controls or receivers, emergency cases etc.

Another consequence of mobile communication is that the driver loses a large part of his autonomy when driving the truck, or stated otherwise that the home-office gains control over the driver. It may be hypothesized that this is in the interest of firms that want to have a structured control over their drivers, for example, because of the lack of informal links. Hence, it seems plausible that large firms are more intensive users of mobile communication technologies.

Finally, autonomous (i.e., without a link with the home office) navigation systems, are clearly of direct benefit for drivers only and not to the home-office, in particular when the driver has to find his way in new situations. Such new situations may relate to an incidentally congested route, or to an entirely new route. Autonomous navigation will therefore most likely be used by firms with predominantly stand-alone contracts, and also by firms which operate on a large spatial scale, since drivers tend to be familiar only with road situations when driving in a relatively limited geographical area.

In any case there is a variety of telematics options for the trucking sector, and it is therefore interesting to identify which options are used for which objectives and under which conditions. The aim of this chapter is to give a comprehensive evaluation of this potential market of telematics systems for road freight transport operators. It presents the results of a Dutch survey among freight transport operators on users and markets of the potential of freight- and fleet-related telematics applications in this sector. Given the factual background that the country of the Netherlands plays an important logistic role in Western Europe as a so-called 'distribution nation' and since it has the highest European share in international freight transport, this is a key country to investigate any potential interest in freight telematics and fleet telematics systems. An assessment has been made in the Netherlands of the investments in telematics equipment, the purposes for which these applications were used, the perceived benefits of and the main decision factors to invest in transport telematics, and the bottlenecks in investment.

The present chapter is structured as follows. Section 8.2 contains information on the survey context and design, followed by Section 8.3 which contains the results of the analysis. In Section 8.4 it is investigated to what extent relevant company characteristics might influence the potential of telematics applications, while Section 8.5 contains the main conclusions and some strategic lessons.

8.2 Survey Design

The survey was conducted in the Netherlands in 1994 and consisted of structured postal questionnaires sent to directors of 1172 road transporters. The data and addresses of the firms were collected from the automated register of the Dutch Association of Chambers of Commerce. The survey addressed all possible categories of goods transport by road. These categories comprise pick-up and delivery services, courier services, transport of light goods and animals, container transports, heavy goods transports, bulk goods transports, trucking services and transport of chemical/dangerous goods.

A statistical subdivision according to the number of employees of the total population of all Dutch road transport companies from the above mentioned categories is given in Table 8.3. It shows that about 80% of the total number is made up of small companies (less than 10 persons employed). However, when looking at the total number of persons employed, the larger companies (more than 20 persons) made up about half of the total employment in the road transport industry.

Table 8.3. Company size (employees) of Dutch road freight transporters (1994).

Number of employees	Number of companies	% of companies
< 5	6566	65%
5-9	1338	13%
10-19	1063	11%
20-49	770	8%
> 49	346	3%

Source: Dutch Association of Chambers of Commerce.

The survey has been stratified (by using a truncated sample, consisting of the upper segment of size categories) in order to include a sufficient number of middle-sized and large companies. These companies make up a relatively high proportion of the freight market in terms of persons employed and operated vehicles, while large companies are usually market leaders whose absence might not have generated sufficiently interesting insights in the developments of the sector concerned. In addition, large firms were expected to be more intensive users of the above technologies and thus more interesting for our investigation.

The survey aimed to investigate the potential of transport telematics applications in the road transport sector from a user perspective. The following subjects are addressed: the actual use, the purposes of use, the perceived benefits, the intentions to invest and the thresholds to investments in transport telematics applications.

The first part of the survey concerned relevant characteristics of road transport companies. These are the number of vehicles, the geographical area in which the company operates, type of freight transported, the relationship with forwarders, and the existing level of automation of the company's trip scheduling and route planning. It was assumed that the first four characteristics determine the budget available for telematics investment and the direct operational benefits to the company. The level of automation of trip and route planning functions indicates the existing experience of a company with technological investments that improve fleet operations. It is to be expected that the potential of transport telematics is highest for proven 'technology-minded' companies.

The second part of the survey investigated the existing level of awareness regarding the telematics applications as distinguished in Section 8.1.

The third part concerned the actual use of telematics systems and the level of satisfaction of the companies that have invested in them. In our analysis, the purposes for which telematics systems are used have been investigated; in the survey, specific functions have been distinguished like the planning of the fleet, the monitoring of dangerous goods etc.

The fourth part of the survey addressed the intentions to invest in telematics or to expand existing investments, including perceived requirements of telematics systems and the magnitude of possible investments, and the final part of the survey investigated the most striking bottlenecks that prevent companies in the road transport sector from making telematics investments.

The response rate of the survey was very encouraging and suggested at least the high level of interest among road transport companies in telematics issues. 325 companies filled in the questionnaire and mailed it back. From these, 320 were useful, which means that a net response of 27% was achieved. Given the fact that commercial freight operators in the Netherlands have recently faced a high number of postal surveys in this industry, this is indeed a relatively high response rate for a survey carried out in this 'over-surveyed' sector, and may be an indication for the perceived importance of these new products.

8.3 Profile of Companies

Since the survey has been targeted to the upper segment of companies concerning their size in terms of persons employed, the profile of responding companies followed a pattern different from the national figures, and was hence biased to categories of firms with 21 and more vehicles (Figure 8.1). The relative shares of the size categories of all firms operating 21 and more vehicles categories were however comparable to those categories of the national population. The categories 21-30, 31-40 and more than 40 vehicles made up about 25%, 10% and 25% of the survey, respectively.

The geographical area in which the firms operated is concentrated in Western Europe (60% of the companies), while in total 70% was internationally oriented (Figure 8.2). Only 25% of the sample appeared to operate regionally or nationally.

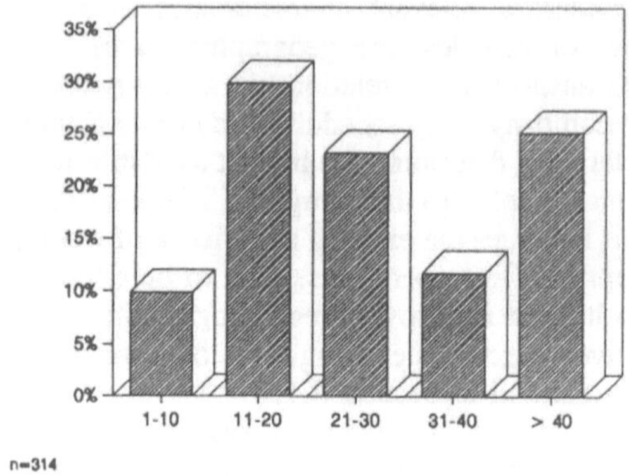

Fig. 8.1. Vehicle fleet size of respondents (% of total sample).

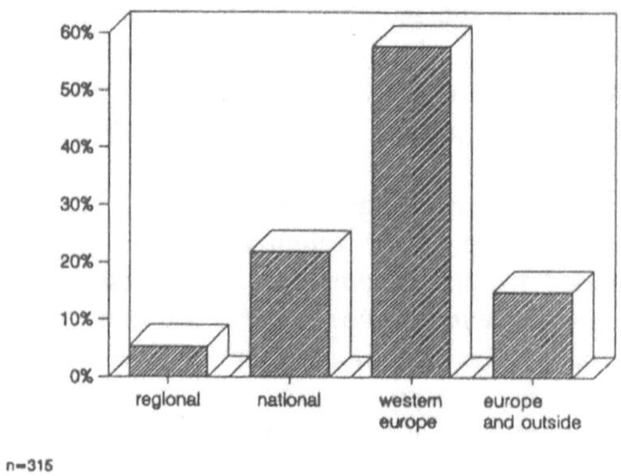

Fig. 8.2. Geographical area of fleet operation (% of total sample).

About 55% of the companies transported particular freight types. Transporters of low-valued heavy goods, high-valued light goods and perishable goods made up about 25%, 15% and 10%, respectively. About 45% of the companies were not specialized in any type of goods.

More than half of the companies claimed to work on the basis of fixed contracts with forwarders, while 10% worked only with incidental orders. This meant that the majority of companies dealt with regular routes and destinations.

The existing level of automation of fleet trip planning and route planning operations was considerable. About 30% of the companies used advanced technology for at least one of both activities: for fleet trip planning, route planning and for both activities about 15%, 5% and 10%, respectively.

8.4 Familiarity with Telematics Systems

Companies had been questioned on their familiarity with a range of telematics systems (Figure 8.3). In general, about 40% of them were not familiar with the possibilities of these technologies, which is a relatively high share. Another 40% were quite familiar with its potential, while only 10% were already very familiar. It seems that there were no large differences between the technologies distinguished in this survey. EDI was in general slightly more familiar to these companies than road telematics systems, while they were slightly less familiar with the possibilities of in-car navigation systems.

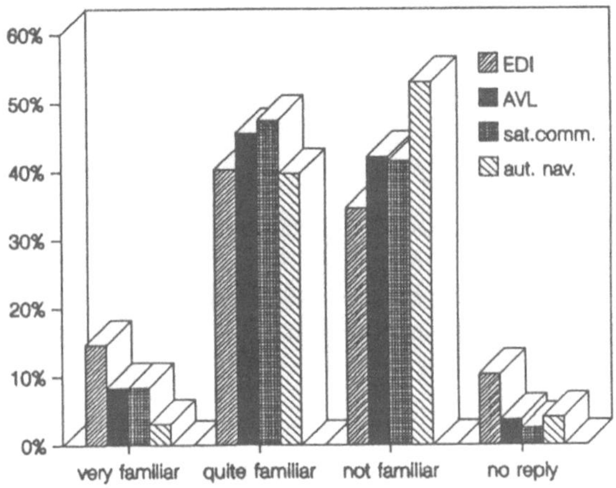

Fig. 8.3. Familiarity with telematics systems (% of total sample).

8.5 Actual Use of Telematics Systems

EDI was currently been used by 48 of the 320 companies (15%). Given the facts that just about a third of the firms are unfamiliar with EDI, and moreover that only 15% of the surveyed companies actually use it, indicates that the EDI market is far being saturated. From the users, 60% used the system as frequent as was expected in the period of purchasing it. About 20% used the system more intensively, while another 20% uses it less intensively than was expected.

Next, users of EDI appeared to be overwhelmingly satisfied with this technology. Almost none of these users were negative: only 2% were unsatisfied and 10% were neutral. About 70% were satisfied and 20% even very satisfied about the implementation of EDI in their company. This appears to be very encouraging for EDI manufacturers, although two remarks are in order here. Firstly, statements about satisfaction of an adopted technology should always be read with care, since entrepreneurs may not be eager to admit that they are disappointed about their innovation and investment. Additional objective information (e.g., the share of EDI in the information processing activities of the firm) is therefore preferable. Secondly, there may be some influence of stratification at this point, because only those firms that currently use EDI responded to this item. In particular, those firms that previously used EDI, but have stopped using it because they were unsatisfied, are badly underreported in this respect. In the same context it may also be indicative that just about 20% of the users is *very* satisfied with EDI.

The number of companies that had already invested in road telematics systems was low. Car (mobilo)phones were relative popular; about 25% had invested in such a mobile communication system. The more advanced technologies (AVL and satellite communication) were applied by respectively 3% and 5% of these companies. No company had invested yet in autonomous navigation systems. It should be noted that the data with respect to satellite communication should be carefully considered. As noted in Section 8.1, during the period of the survey, satellite communication was challenged by the introduction of the GSM mobile telephony, suddenly making the former rather almost redundant. Hence data related to satellite communication in the past have a different context compared to those related to the future. This may complicate the interpretation of the results.

The low adoption rates (except for mobile telephony) stand in contrast with the familiarity with these systems, which is however not unexpected. It means that either firms are not convinced of their usefulness and hence will not use them in the near future, or that there is a large potential market which just has to be developed.

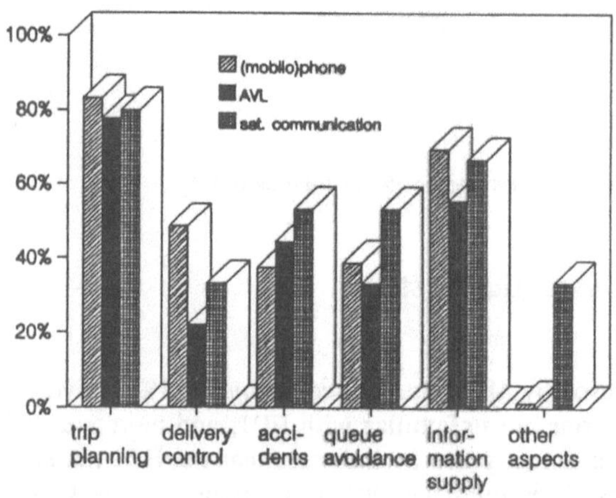

Fig. 8.4. Purposes of use of telematics systems (% of respective user groups).

In Section 8.1 it was argued that AVL is most likely used in combination with mobile communication. This is confirmed by the data, since indeed all nine AVL users also used mobile telephony or satellite communication. On the other hand, mobile telephony (MT) and satellite communication are not considered as substitute systems, since eleven firms reported the use of both technologies. It may be seriously questioned however, whether these firms view the two technologies as distinct, or rather think that MT use is based on satellite technology.

When looking at the purposes of use of those systems currently being applied by the companies (Figure 8.4), it seemed that the main purposes of use were the planning of the trips of the vehicles and the information supply to clients. Less mentioned were purposes like the control of deliveries, coordination in case of accidents and the avoidance of congested roads.

It also seemed that there was a slight difference between mobile communication systems and vehicle location systems. The former are slightly more intensively used for purposes like delivery control, congestion avoidance and information supply to clients.

The levels of satisfaction with road telematics systems were, just as in the case of EDI, quite high. Only in the case of car (mobilo)phone, a share of 7% was unsatisfied. In the cases of AVL and satellite communication, there are no negative responses, while more than 60% were satisfied and 40% was very satisfied. It must be noticed here that these results are based on only a limited number of actual users, which make general conclusions difficult to infer.

8.6 Investment Plans and Barriers

All respondents had been asked about their plans to invest (or to expand existing investments) in EDI within a (limited) period of two years. About 45% of them had no plans to invest, while 30% were uncertain. Another 20% claimed to have serious plans to invest in EDI within two years.

Those who were negative to any investment were asked about the most striking or significant reasons (Figure 8.5). It appeared that two of the most important reasons were the lack of needs of clients (65%) and the preference to wait for further technical EDI developments (60%). A third reason was the lack of EDI investments by clients (40%). Less mentioned motives were size of the company and the high investment costs.

Existing plans to invest in the advanced types of road telematics systems (note that investment plans in conventional mobile telephony was not surveyed) were less than in the case of EDI, but nevertheless considerable. About 63% of the companies responding to this question explicitly stated to have no serious intentions to invest in either of these systems within a period of two years, while another 13% were uncertain about it. In the separate cases of each of these systems, 3% to 13% had serious plans. The share of positive answers was the highest in the case of satellite communication systems (13%), followed by AVL (7%) and autonomous navigation equipment (only 3%).

It was the majority of the current AVL and satellite communication users who had plans for further investments in the applications. It was remarkable that almost all AVL users had plans to invest in satellite communication, but that satellite communication users were more hesitant to invest in AVL. Moreover, firms tended to combine technologies (see Figure 8.6). In Figure 8.6, each subset consists of those respondents with a positive willingness to invest in any of the technologies, and the intersections of the subsets thus show the number of firms that are indicating to invest in both, or even all three technologies. It appears that firms in general have intentions to adopt technologies in combination, apparently expecting synergy effects. This holds however, to a lesser extent for satellite communication, which apparently is also regarded as an interesting technology on its own. AVL seems thus perceived as particularly useful with (advanced) telecommunications possibilities, whereas these systems can later be used as stand-alone devices.

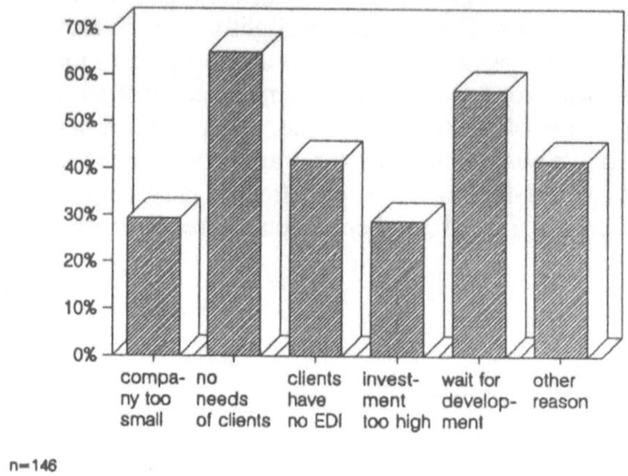

Fig. 8.5. Mentioned reasons not to invest in EDI (% of negatives to EDI investment).

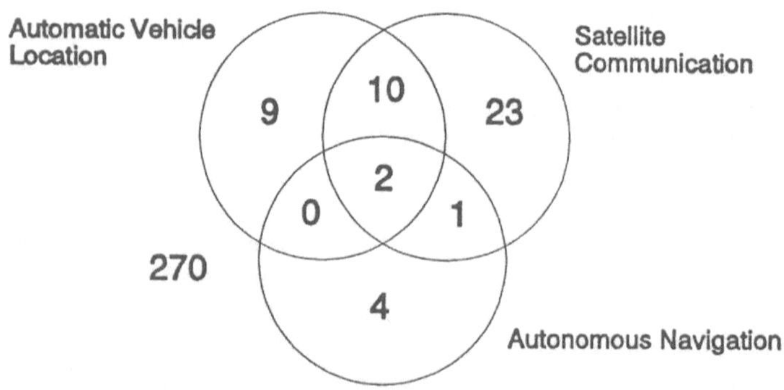

Fig. 8.6. Number of firms with a positive willingness to invest in advanced road telematics systems.

The companies that were not negative to any investment in either of these systems were asked about the advantages they perceived to be able to generate from the use of these technologies (Figure 8.7). In answering this question, firms were not asked to specify which specific reason was related to which technology (in case they had investment plans for more than one technology, which is commonly occurring). The possible benefits, to which the respondents mainly adhered as could be expected, can be combined in three groups with increasing degree of concreteness (see also Ouwersloot et al. 1995b): strategic aspects, concrete qualitative improvements - information supply to forwarders and receivers, speed and punctuality of delivery times, control traffic congestion, coordination at trouble /monitoring dangerous goods - and contribution to cost reduction - efficiency/load ratios of vehicles, reduction driven kilometres.

The advantage which was most often mentioned was the possibility to improve information streams to the forwarders and receivers, an issue which was mentioned by 75%. This was followed by improvements in the accuracy of delivery times and the efficiency and mileage of the vehicles (all about 55%). Less mentioned issues were the company's strategic position, the avoidance of traffic congestion and the monitoring of dangerous goods.

Firms thus tend to concentrate on the qualitative and cost aspects. The number of firms mentioning strategic benefits explicitly is substantial, but they always mention at least one of the other benefits too. Indeed, these latter two groups of potential benefits are hardly ever mentioned alone as the reason to consider new investments.

Table 8.4 relates the familiarity with the technology to the perceived benefits. The table shows that familiarity leads to higher expectations concerning the strategic and qualitative benefits, but shows no impact on the expected cost benefit. Also notice that the absolute percentages for each potential benefit do not differ dramatically between familiar and non-familiar firms, and also not across technologies, and this observation is particularly for the qualitative and cost benefits. Hence, it can be concluded that familiarity leads to increased expectations concerning less concrete benefits, and also that for each of the technologies the benefits are perceived as similar.

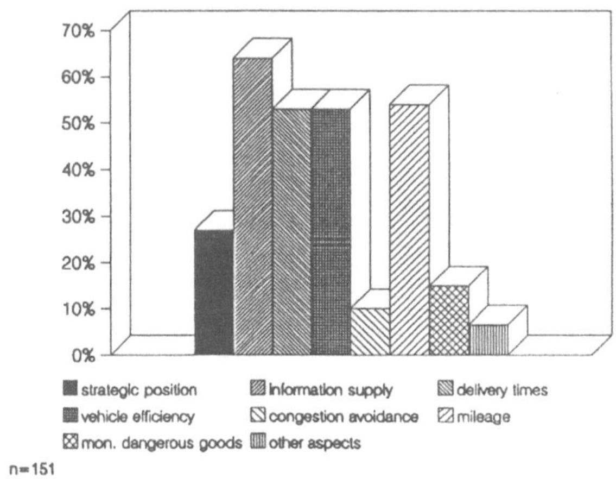

Fig. 8.7. Perceived benefits from road telematics (% of positives toward investment).

A very important issue was the price of the equipment which would be considered to be acceptable in the cases of the three technologies (AVL, satellite communication and autonomous navigation - Figure 8.8). Maximum acceptable prices for the respective on-board equipment seem to be relatively higher in the case of satellite communication systems. In the case of the other two technologies, only about 3% of the companies that were not negative to investments in the respective technology claimed to be prepared to pay a price of more than 3250 ECU for the on-board equipment. For satellite communication, this figure was 8%.

Table 8.4. Relation between familiarity and types of perceived benefits (in percentages).

	Familiarity with respective technologies					
	AVL		Satellite comm.		Auton. navigation	
Benefits	Very/Quite	Not	Very/Quite	Not	Very/Quite	Not
Strategic	37%	24%	34%	30%	39%	29%
Qualitative	92%	88%	93%	81%	95%	86%
Cost	82%	88%	86%	81%	84%	86%

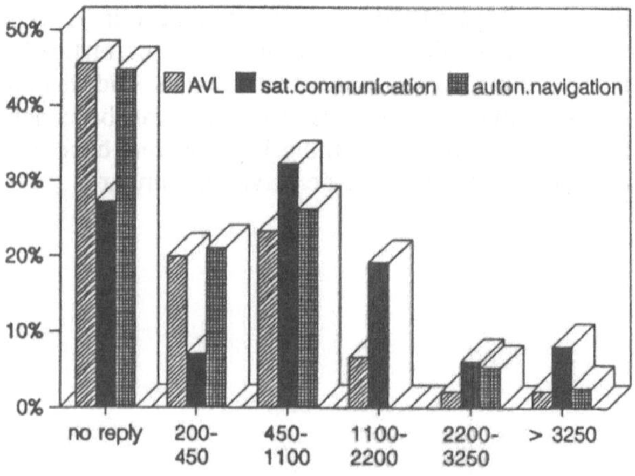

Fig. 8.8. Perceived maximum investment costs of on-board equipment (ECU) (% of firms with a positive attitude toward investment).

For the related office and management equipment, only 5% of those companies that were not negative to investments in at least one of the above mentioned road telematics systems stated to be prepared to invest more than 21,000 ECU.

Next, we will concentrate on the difference between firms which are respectively users and non-users of specific technologies. It seems reasonable to expect that intended investments are positively related to the degree of familiarity with the technology concerned. The data show that this indeed holds for AVL and satellite communication. This familiarity with the technologies also explains the relatively large difference in indicated investment amounts per vehicle, the firms are willing to pay. That is, non-users that have investment plans typically report much lower amounts. For example, none of the users indicated a maximum amount per vehicle below 450 ECU, whereas 10% (AVL) and 6% (satellite communication) of the non-users are in this range. This may be interpreted as a clear under estimation of the required investments in the technology due to unfamiliarity. However, the observation made when discussing the investment plans for users that satellite communication is perceived as more expensive is confirmed, as the average amount for this technology is expected to be higher. Moreover, the willingness to pay for autonomous navigation systems appears to be in the same order of magnitude as for AVL. This suggests that these two technologies are perceived as equally useful.

Another aspect which may be related to the level of intended investments are the perceived benefits. An *ad hoc* reasoning would suggest that the more benefits a firm perceives, the higher the intended investments, and also it seems plausible that the more concrete these benefits are, the higher the investments. However, applying a linear regression to explain the maximum amount per vehicle that the firm is willing to invest with the three benefit variables defined above reveals that these benefits play no significant role whatsoever. On the contrary, the fact that a firm is already a user of the respective technologies is by far the most important factor explaining differences in the level of the investment plans. This is quite surprising, since it was also observed that the familiarity with the relevant technologies made firms somewhat sceptic about the expected benefits. Apparently, firm-specific and site-specific experience leads them to higher investments.

The picture that emerges from this discussion is that firms do not have a clear picture of what benefits telematics applications can bring to them. They suspect there to be benefits, but are unable to indicate clearly where they can be found. Consequently, the answers to the investments lack any structure, and are presumably driven by other considerations than a cost-benefit evaluation.

The overall suggestion that comes to the fore is that firms which have invested in telematics technology start to learn about it, experience that it is not so cost reducing as they had thought, but that the less concrete benefits are more real, and therefore inspire them to raise the investment levels. Another aspect that will contribute to this explanation is that experienced firms have better (i.e. higher) cost estimates of the investments required than uninformed firms.

All respondents were asked about the bottlenecks they perceived as most striking regarding any investments in AVL and satellite communication technologies (Figure 8.9). The most frequently occurring bottleneck seemed to be the costs associated with the implementation of the systems (about 45% of the companies mentioned this aspect). A second main bottleneck was the uncertainty about the speed of development of these systems, mentioned by about 35% of the companies. These are followed by issues like a too small area in which the company operates to generate advantages from the use of these systems and the fact that these advantages are still unclear to the companies (both bottlenecks mentioned by about 25%). Issues that seemed to be of less relevance were the size of the vehicle fleet (15%), the perception that the available infrastructure of the systems is not yet adequate and detailed enough, and the quality of the staff to work with these systems (both less than 5%).

All of the respondents were asked about the perceived bottlenecks to implement an autonomous navigation system to their fleet (Figure 8.10). The most important issue was the fact that drivers are familiar with the routes they use to drive (mentioned by about 50% of the respondents). This issue is followed by maps providing sufficient information to the drivers, uncertainty about the developments of these systems and too high investment costs (all mentioned by about 30% of the firms). Slightly less important issues are here the area size in which the firm operates and the ambiguity of advantages (both about 20%).

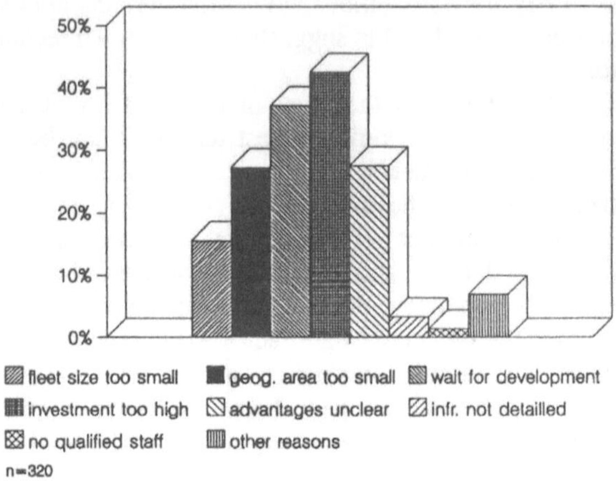

Fig. 8.9. Reasons not to invest in AVL/satellite communication (% of total sample).

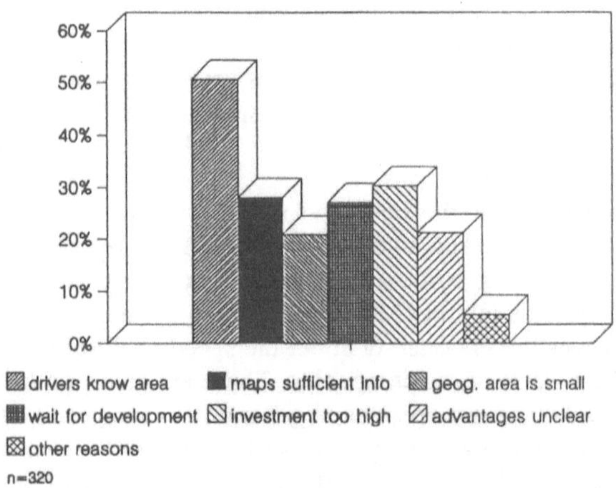

Fig. 8.10. Reasons not to invest in autonomous navigation (% of total sample).

8.7 Dynamics in Adoption of Systems

The results from the above survey provided an opportunity to compare the attitudes within the road freight sector toward telematics with the situation of three years ago, on the basis of a similar survey carried out by the Dutch Ministry of Transport in 1992 (Rijkswaterstaat 1992b) on the potential of AVL technologies in particular. However, such a comparison should be carried out only with taking into account the differences which exist between the two surveys concerning inter alia the sample selection and the data analysis. Especially caution is warranted with regard to the sample composition, as the earlier survey was not stratified for larger companies. The conclusions are summarized below:

Familiarity with systems:
* Three years ago there was insufficient familiarity with the possibilities of AVL systems. For example, only 2% of the companies expressed to be familiar with the LORAN-C system. This awareness seems to have slightly increased. In 1994 only 40% of the companies declared not to know AVL technologies.

Use of systems:
* Three years ago almost none (1%) of the Dutch road transporters used EDI. This has changed rapidly; about 15% of the companies use EDI in 1994. This indicates a rapid development regarding the implementation of EDI.
* Three years ago only about 2% had invested in an AVL system. This figure has not changed significantly: in our survey only 3% used an AVL system. It may thus be concluded that the process of EDI implementation is more dynamic than AVL implementation.

Investment plans:
* The number of companies having intentions to invest in EDI seems to have changed considerably also. While three years ago the number of companies having intentions to invest in EDI within a period of 5 years was about 10%, in 1994 20% of the companies were willing to invest within the next two years. One has to take into account however, that the survey in 1994 has been stratified to the larger companies, while this was not the case with the one in 1992.
* The existing intentions to invest in road telematics systems were still low. Positive intentions to invest in AVL were expressed by 7% of the companies, while 25% were unsure. Three years earlier similar measures showed respectively 8% and 8%. Only an indicative conclusion may be drawn, namely that there are only slightly more companies nowadays considering the possibility to invest in AVL systems.
* The most frequently mentioned arguments against investments in road telematics have not changed in the past few years. The most important ones are still the perception that drivers are in most cases familiar with the area in which they drive, and the size of investment costs.
* More than half of the companies with plans to invest in AVL systems intended to combine this with satellite communication. From the earlier survey, it appeared that all of these companies wanted to combine AVL with any mobile communication

system. This fact stresses the need for a product combining both functions.

8.8 Influence of Company Characteristics

The influence of some typical key company characteristics, such as the number of vehicles operated, the geographical area in which the company operates, the type of freight transported, the relationship with forwarders, and the existing level of automation of the company's trip scheduling and route planning were also investigated. It was hypothesized that the first four characteristics determine the budget available for investments in telematics and the direct operational benefits to the company, while the level of automation of trip and route planning functions indicates the existing experience of a company with logistic investments in fleet operations. It may be assumed that the potential of telematics is highest for proven 'technology-minded' companies. Contingency table analysis has been used to identify relationships showing a sufficient significance level (below 5% using the Pearson method) which were further analyzed by a logit analysis.

Use of systems

The actual use of EDI was positively correlated with the size of companies. Small and medium-sized firms showed a lower use of EDI than big companies (Figure 8.11). From the companies operating less than 20 vehicles, 7% used EDI, while for companies operating more than 40 vehicles this figure was 30%. In the case of mobile telephony, these respective figures are 20% and 35%, which means that this product is relatively more popular in the smaller categories compared with EDI. In the case of AVL these figures were 1% and 8%, and in the case of satellite communication 2% and 10%, respectively. It must be remembered, that in the case of the latter two products the absolute numbers are small, so that the results may not be representative. These descriptive figures nevertheless indicate that the largest size category of firms indeed plays a pioneering role in the case of all technologies.

The relatively high adoption rate of EDI in the sample allowed to apply a more advanced explanatory statistical investigation of this advanced technology. From the description in Section 8.1 we expect that in addition to large firms (with reference to conventional diffusion theory), EDI will be more often used by firms with predominantly fixed contracts, opposed to stand-alone contracts, and by firms facing more control points, i.e. international transport companies. These hypotheses were investigated by estimating a logit model, with EDI adoption as the endogenous variable, and the above mentioned variables as the explanatory variables (see Table 8.5). The results show that the dominant type of contract has the a priori expected effect on EDI adoption. This result stresses the importance of the factor of uncertainty which companies face with regard the use of EDI by their relations. This uncertainty might be lower in a situation with fixed relations. Also, world-wide operating firms are more frequently users of EDI, as well as firms from the largest size category. However, estimates for the other categories are less in line with our prior expectations, but it should be recognized that these estimates are also not significant. So in general we may conclude that reasonable support for our hypotheses is found in this estimation.

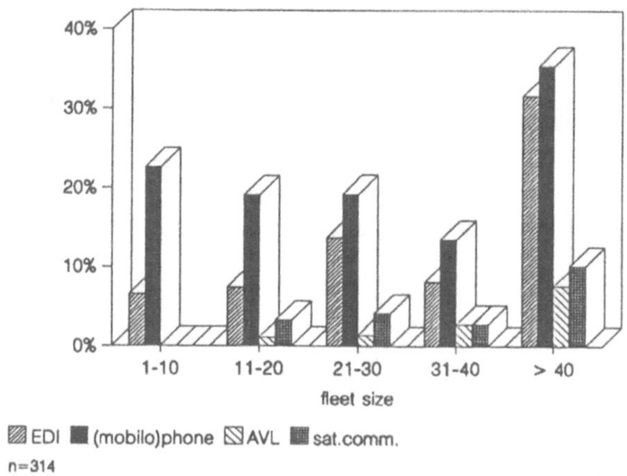

Fig. 8.11. Use of telematics systems for size categories.

Table 8.5. Logit estimation of EDI adoption.

Variable	FIXED	ST_ALONE	NATIONAL	W_EUROPE	WORLD	SIZE2	SIZE3	SIZE4	SIZE5	CONSTANT
Estimate	1.01	-0.25	0.61	-0.16	1.22	-0.09	0.77	-0.09	1.77	-3.42
S.E.	0.40*	0.83	1.15	1.14	1.17	0.88	0.86	0.99	0.84*	1.19*

-2** Log Likelihood 264.79 constant only
-2** Log Likelihood 221.97 full model
Model Chi Square 42.82*

* significant at 5% level ** significant at 10% level

The use of AVL and mobile communication systems was finally also checked against the geographical area in which the firms were operating. It appeared that there was a strong relationship between use of these and the geographical orientation. In the case of AVL, had 4% of the internationally operating firms in the whole sample invested in it, compared with 1% of the companies with a regional or national orientation. In the case of satellite communication these figures were respectively 7% and 0% (care has again to be taken with regard to the low numbers of AVL and satellite communication users in the sample). This did not account for mobile telephony, whose use was quite equally spread over the geographical orientation of firms.

Investment plans

Given the emerging influence of some company characteristics on the current adoption of EDI by road freight transporters, it was also meaningful to investigate the determinants of plans for future EDI investments. Again we have used a logit model to investigate these determinants. The explanatory variables are the same as in the estimation of the use of EDI, but in addition the use of EDI was also included here. The results are given in Table 8.6. Clearly, the current use of EDI and the existence of investment plans are strongly related. Apparently, firms are truly satisfied with EDI and therefore foresee an expansion of its use. The other explanatory variables tell more or less similar stories as in Table 8.5, but with a different emphasis. The spatial scope of the firm's activities appears to be a reasonable explanatory variable, at least supporting the hypothesis that internationally operating firms are relatively more interested in EDI. Also the size of the firm gives the a priori expected result. The dominant type of contract however, does not lead to the expected outcomes, which is less easy to understand. A possible - but speculative - reason may be that fixed contract firms have reached their saturation level of EDI investments in contrast to other groups, as they were the first to adopt the technology, and may also be expected to be the first in this sector to complete their equipment.

Table 8.6. Logit estimation of EDI investment plans.

Variable	EDIUSE	FIXED	ST_ALONE	NATIONAL	W_EUROPE	WORLD	SIZE2	SIZE3	SIZE4	SIZE5	CONSTANT
Estimate	2.39	0.12	0.67	0.45	0.42	0.74	-0.44	0.68	0.68	1.93	-3.24
S.E.	0.41*	0.38	0.57	0.98	0.95	1.01	0.86	0.81	0.87	0.78*	0.90*

-2** Log Likelihood	325.55 constant only
-2** Log Likelihood	231.69 full model
Likelihood ratio test	93.86*

* significant at 5% level   ** significant at 10% level

It is interesting to remark at this stage that it is really the actual use of EDI that counts in explaining investment plans, not the reported level of satisfaction. When this latter variable is used (by means of two dummy variables, one for firms which are very satisfied, and a second one for 'normally' satisfied firms), instead of the mere EDI use, the estimation leads to statistically inferior results, even despite the increased number of explanatory variables!

The (tentative) conclusion that can be drawn from this previous analysis is that firms tend to adopt EDI in the predicted way (i.e., they appear to adopt it rationally) and that the adopters are almost unequivocally satisfied with the technology, even to such extent that users have a strong incentive to make further investments. Finally, the analysis suggests that firms working predominantly with fixed contracts have adopted EDI first, but also have completed their investments earlier.

In the case of investment plans in AVL and satellite communications, also a positive relationship exists between the willingness to invest and the size of the company (Figure 8.12). In the case of AVL systems, there is an increasing share of companies for the successive size categories showing a positive attitude to investments within the next two years (from 0% of companies with less than 10 vehicles to 14% of the large companies with more than 40 vehicles). In the case of satellite communication systems these figures appear to range from 4% to 22%.

When relating these investment plans with the geographical area in which companies were operating, it becomes clear that those companies operating internationally tend to be more positive towards future investments. In the case of AVL, 9% of the internationally operating firms in the whole sample had plans to invest in it, compared with 4% of the regionally and nationally oriented firms. In the case of satellite communication, these figures were respectively 16% and 3%.

Investment plans in AVL and satellite communication apparently have no relationship with the existing level of automation of company functions that assist trip and route planning. This means that interests in road telematics systems is expressed to a similar extent by companies that did not use computer-assisted logistic planning of their trips and routes as by companies that managed logistic planning assisted by computer software applications.

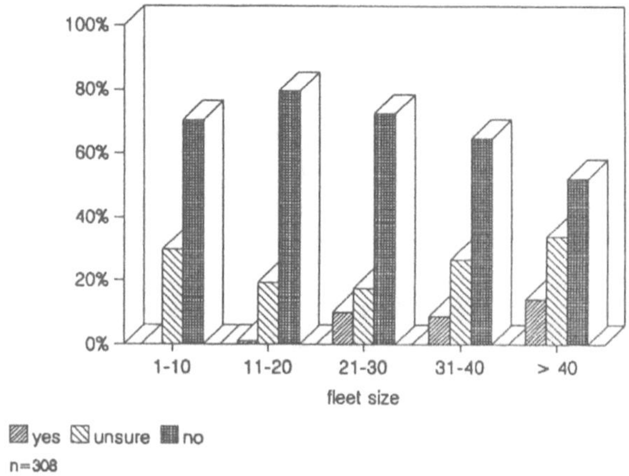

Fig. 8.12. Investment plans in AVL systems for size categories.

Finally, the relationship between investment intentions in AVL/satellite communication systems and the type of freight specialization has been investigated. It might be expected that dynamic information on freight movements would be the more useful the more the freight has an urgent character. It indeed appears that companies specialized in very urgent goods showed intentions to invest to a larger extent than companies specialized in heavy freights. This is clearly the case for both technologies AVL and satellite communication.

8.9 Conclusions on the Telematics Market for Road Freight Operators

Adoption and successful market penetration of new technologies follows normally a logistic growth curve, starting with a hesitant use in initial stages followed by accelerated market coverage up to a stage of gradual saturation. The trajectory of telematics system in the commercial trucking sector seems to follow largely this standard picture. Consequently, there are - as usual - early adopters and late adapters. The survey among fleet operators presented in this chapter has revealed the nature of the various transport firms in terms of their willingness to use telematics as an integral part of their activities.

In the empirical analysis a distinction has been made between EDI and other telematics applications (notably mobile telephony, automatic vehicle location, satellite communication and on-board navigation). As far as EDI is concerned, it appears that firm size plays a dominant role in the use of EDI, while next the nature of contracts of carriers (in particular, fixed versus stand-alone contracts) and their international orientation are other major explanatory factors.

For the other types of telematics systems, it appears again that firm size is a critical variable. Familiarity with new telematics systems and investment plans are not uniformly distributed among firms, but are related to the perceived or expected benefits of telematics.

Based on such findings, one may try to draw inferences for the future trajectory of telematics applications in the freight transport sector. First, telematics use is still in its infancy and there is a wide scope for a much broader market coverage. Second, the market of telematics users is segmented, so that it seems plausible that telematics applications will have various appearances, depending on specific firm needs and characteristics.

For EDI, it appears that so far the critical mass for its use has not yet been reached, so that one may expect that this network technology will generate a much higher share in all segments, as soon as the externalities of the technology lead to a significant rise in the benefits for all actors involved.

For the other telematics technologies such network considerations are not relevant. They may simply be assumed to follow the typical logistic adoption curve applicable to innovations in general. The results of the present analysis do not point to any other adoption patterns as the most fundamental hypothesis involved in the general diffusion theory - large firms adopt first - is confirmed. An exception however, should be made for satellite communications systems. For this technology, take-off has been particularly slow, and its technical advantages has been caught up by simple mobile telephony - in particular by the introduction of GSM - which is also much cheaper. Hence we suggest that satellite communication systems as applied in the road freight transport sector may have a more limited future than has been predicted.

PART C

PERSPECTIVES FOR TRANSPORT TELEMATICS

9 Views of Public Authorities on Telematics

9.1 Introduction

This chapter will deal with the role of public authorities in the implementation process of road transport telematics. Basically, this role has a two-fold character. First, there is the role at the strategic level, where the public sector has its impact on the direction of the development of the technologies. Second, there is the role at the operational level. It is here where road authorities (in most cases public sector departments) are responsible for the direct guidance of the developments in overall traffic conditions. The roles of the suppliers of the telematics systems have been presented in Chapter 3. But the public authorities also have a major responsibility for the research and development necessary to promote telematics and in the adoption of European standards for implementation.

9.2 Goals of Public Authorities

The intervention of public authorities in the implementation process of transport telematics, which are mainly road management institutions associated with departments of local and national governments, is based on a variety of objectives. A few of these objectives can be identified as the main reasons for this intervention (O'Neill 1990). They are directly related to the three main potential benefits of transport telematics which were identified in Chapter 1.

The first reason is related to traffic safety. The technical developments of vehicles and roads has been directed towards reducing both the probability of accidents and their severity. Transport telematics has in recent years had a major role to play, and this requires laws and regulations to prescribe requirements for the design and construction of vehicles and their equipment. These are clearly governmental tasks at the national and international levels.

In parallel with safety considerations there are a range of environmental considerations, where the authorities' policies have been aimed at mitigating forms of physical harm (to human beings, properties and the natural environment) and psychological nuisance. In this respect may be mentioned visible external effects of traffic: noise and vibration, air pollution, visual intrusion of roads and vehicles, etc. In addition to new engineering means, transport telematics technologies can potentially make a major contribution to the reduction of these negative environmental traffic impacts. Because these mentioned effects go beyond the individual level and are of concern to the whole society, there are clear responsibilities imposed on the public authorities regarding the development and implementation of transport telematics applications.

Finally, public authorities have the general objective of promoting economic efficiency, and inherently the efficiency of the transport system. Transport inefficiency has negative impacts on individual travel time and incurs economic losses to regional and national economies, above all on congested roads. As such, public authority institutions have the responsibility for the performance and guidance of traffic flows on urban roads and inter-urban road corridors. Transport telematics may help individuals to minimize their own and other road users' losses of time, it will help to expedite traffic flows along these urban roads and inter-urban corridors, subject to necessary constraints, and it will provide the means for market pricing where this is chosen and approved by public authorities. Their role is even more important given the consideration that for many transport telematics applications, the efficiency of individual trips may be increased, although this will not necessarily result in a greater overall benefit to the transport system in terms of safety, traffic flow efficiency, or a better quality of the environment (Emmerink et al. 1994a).

In parallel to intervening in the transport telematics market for the sake of the public, social and economic interest in the performance of road transport, road authorities have also a major interest in the supply of transport telematics equipment and services. In this new market there is strong competition both between domestic transport telematics industries and at the international level within the European, North American and Pacific regions, implying a pressure on public authorities to favour domestic transport telematics industries.

It is thus clear that in investing in and implementing transport telematics equipments, road authorities start usually from a firm strategic policy basis for transport in the area they are concerned with. These policy objectives are then developed into a number of operational strategies some of which may be applicable and relevant to transport telematics products. Each strategy has various methods of implementation, and transport telematics methods will have to be compared to alternatives in order to evaluate the most cost-effective way of achieving the strategy, and the policy objectives concerned. Road authorities need to demonstrate a significant added value accruing from costly transport telematics investments. In this context, the potential of road transport telematics from the perspective of these road authorities is investigated in the next section.

9.3 Public Authority Interventions

Interventions of public authorities in the transport telematics implementation process may take place at different operational levels. A distinction is made between the local/regional level and the national level. Each level has its specific operational goals and responsibilities with regard to transport telematics (O'Neill 1990).

9.3.1 The Local/Regional Level

Local public authorities are concerned with the regulation of urban and regional road traffic and with the management of parking space. They cooperate in these functions with national road authorities in the case of areas that are traversed by roads of national importance and therefore financed wholly or partially at higher levels. At the local level, authorities are responsible for the provision and control of parking-space both on- and off-street, and with traffic engineering and control over the local road network, which includes priority management and control of junctions. These include both isolated junctions and those within integrated area traffic control networks, which may include motorways around urban agglomerations and their access roads (e.g. traffic control on urban ring-roads).

Transport telematics applications will require adjustments in the organization of traffic management. The improved collection, processing and dissemination of information about traffic conditions will lead to the integration over wider areas of the provision of advice or instructions to drivers. This may be a sensitive issue, because authorities at the different levels have their specific policies for encouraging drivers to follow certain routes. Local and regional authorities often have the objective to discourage use by through traffic of streets where traffic is accompanied with high local annoyance (e.g. residential or major commercial streets), even though this issue is not a major driving force in traffic policy and management at the national level. In order to maintain a consistent and generally acceptable local traffic policy, local authorities use a series of instruments to deal with such traffic: these may facilitate flows (e.g. by one-way streets) or may discourage traffic (e.g. by imposing restrictions on classes of vehicles, journey purposes, and time and day). Local authorities may be particularly concerned that individual route-guidance services should not be in conflict with the purpose of such regulations, nor direct traffic along other streets with high local annoyance of traffic. The transport related priorities also have to be balanced against other citywide priorities, such as maintaining employment, attracting new industries, housing allocations and improving quality of the environment. Cities have to be seen as attractive places in which to live and work, and transport, including telematics, has an important role to play.

9.3.2 The National Level

Local and regional authorities play an important role in the implementation process of transport telematics, but are dependent upon their national government to provide enabling powers and make laws and regulations, as well as exercising overall control of the use of nationally levied taxes. Authorities at the national level set the framework for decisions at the local level by weighing up overall national interests and international constraints.

One aspect of growing concern to the national government is the question whether road pricing should be introduced as a means of regulating traffic, and/or of financing road infrastructure. Transport telematics technologies provide the possibility of applying road pricing on inter-urban road corridors and also on motorways in and around cities, where most traffic congestion is concentrated. National authorities have to deal with the socio-economic dilemmas related to road pricing, for example, the equality of access rights and the protection of privacy of motorists. There is in general a public resistance to both the fundamental concept of road pricing as well as to the possibility of central authorities to trace individual's movements through the exchange of information between car and roadside beacons or other detectors (Banister 1994b). This raises a host of practical problems. For example, toll operators would prefer to debit accounts of authorized credit users, and police organizations would not want to be deprived of the ability to track suspected vehicles. While electronic road pricing designers leave the option open that users withhold their identity, it is the task of the national government to choose a particular debiting principle and the related technological options.

Another aspect of particular responsibility for the national government is the issue of traffic safety. Telematics applications for car drivers could in certain circumstances, and depending on the type of interface, detract from their driving task. There are well established procedures for seeking international consensus on safety aspects of road vehicle design and construction. However, there are new problems, both in ensuring the integrity of new systems where current rules for protection from electromagnetic interference may not wholly suffice, and in the critical area of the man-machine interface. Authorities have to decide whether to allow the use of systems such as those for intelligent cruise control. Car drivers expect certain protection from the government and hold it culpable in the event of accidents, especially in the case of the development of cooperative driving technologies.

In general, national authorities are concerned with technical standards for telematics technologies. They may prefer these to be in harmonization with international standards, or they may issue their own requirements as either regulators or purchasers of telematics. In addition, governments have responsibilities to industry and users as they are democratically accountable through the electoral process. Even though congestion, safety and environment are perceived as important issues among the electorate, positive actions are difficult to take, particularly if prices are raised. Industry and users are suspicious that any price rise is in fact a means to raise additional revenues for the exchequer and they will not benefit directly. There seems to be a common view, particularly in the transport sector that the price is already high and that any additional costs must be seen to be reinvested in the transport system in the form of new infrastructure. So, although there are concerns over issues of congestion, safety and the environment, that concern has not manifest itself through the desire for positive action on road pricing. There is a general resentment against new taxes or even paying for what has been considered a "free good" (e.g. road space), at least at the point of use. Governments have a major task in convincing the electorate of the case for road pricing and the use of telematics. So, action here is required both on the technological front to establish technical standards and on the public acceptability front to convince the electorate of the benefits of using the full range of transport telematics systems (Banister 1994b).

9.4 Attitudes and Expectations of Road Authorities: A Dutch Case Study

The attitudes and expectations of road managers towards the possibilities of applying these information and telecommunication technologies to their networks have been explored by a series of in-depth interviews, which were held in the Netherlands with key persons of Rijkswaterstaat, the national public road manager in the country. These interviews were held in June and July 1994. Respondents were selected both at the central department as well as at some regional subdivisions of this state agency. The selection ensures that opinions are included of those experts who are involved in planning and research (central department), as well as of those involved in the actual implementation of new informatics technologies (regional subdivisions).

The aim of these interviews was to assess and understand the (individual) attitudes, perceptions and expectations of inter-urban transport decision makers with regard to advanced transport telematics technologies applied to motorways. These technologies comprise debiting systems, road-based dynamic information systems (like signalling, fog warning, route information) and in-car driver information systems, with a focus clearly on the provision of various kinds of travel information to motorway drivers.

For the interviews a pre-determined list of probes was used. These probes are the following:

1. perceived potential contribution of transport telematics to the goals of efficiency, safety and environment
2. transport telematics and modal split
3. quality of existing driver information means
4. relative role of transport telematics in the range of traffic management tools
5. optimal types of information and ways to present information
6. basic requirements
7. drawbacks
8. relationship between road managers and transport telematics industry
9. bottlenecks to the implementation process
10. parallel existence of various systems
11. differences in the potential of telematics between different regions
12. implementation prospects.

In the following section, the opinions of the respondents on these issues are reflected in the order given above.

Perceived potential contribution of transport telematics to the goals of efficiency, safety and environment

Although it was perceived that the contribution of transport telematics applications to the achievement of higher traffic efficiency will be limited, it was in general believed that this effect will still be significant. The significance of the impact is supported by the fact that traffic queues usually occur when traffic demand marginally exceeds the road infrastructure capacity. This would mean that even when only a small share of drivers ere rerouted, this could already have a large impact on traffic performance. Improvements of traffic flows as estimated by the respondents varied between 3% and 15%.

Opinions on the possible impacts on traffic pollution were more varied and slightly more negative. It was felt that positive impacts on traffic performance would not directly mean that also positive impacts on pollution by cars could also be expected. This effect could indeed be negative since transport telematics policy in fact enlarges usable capacity. By rerouting of vehicles, the total mileage could be enlarged. This effect could neutralize the benefit by reduction of congestion pollution.

Transport telematics has also much potential to improve traffic safety. Some clear examples were mentioned like fog warning systems operational in the Southern part of the Netherlands and signalling systems operational in the Randstad. Fog warning systems appear to be effective; they may reduce the number of accidents caused by fog situations by 15% to 25%. Signalling systems may filter out extreme speeds and smooth traffic, having a positive effect on front-tail collisions (reductions of 50% in secondary accidents were mentioned). Furthermore, the provision of dynamic traffic information to car drivers is thought to have a positive contribution to safety, since it may decrease the driver's uncertainty resulting in less distraction from the driver's task. In general the respondents were optimistic about the impact of transport telematics systems on safety.

Some worries were expressed however about the use of in-car graphic interfaces for presenting travel information to drivers. It was felt that more research should be needed to get more insight in the impacts of the use of these kinds of interfaces on safety.

It was recognized that the potential effect of transport telematics will basically depend on the drivers' response. It was stressed that an absolute condition for the success of transport telematics is that reliable and non-conflicting information is provided, and that the requirements of the end-users are fully met. This has not been achieved so far in the past. Severe doubts existed about the mentality of car drivers. Lack of individual willingness to adhere to driving advice in situations when the road user perceives no direct individual benefit, would retain any potential benefit of transport telematics for the community.

Transport telematics and modal split

Severe doubts were expressed concerning the usefulness of combining driver information with public transport information in integrated telematics devices (e.g. route guidance to park & ride sites). There was no belief in a significant increase in the use of public transport by car drivers. There would be a psychological resistance to step out of the car into public transport on one single trip. Furthermore, the combination of driver information with real-time public transport departure times might result in undesirable situations in which drivers speed up to catch a planned departure time. Safety could thus be decreased with a broad implementation of such integrated information systems.

The only way to increase the use of public transport among road users by better information would be to provide public transport information at home, but also it was doubted that this would have any effect on the use of public transport.

Quality of existing driver information means

Respondents agreed that driver information systems presently available are useful to a limited extent, but their quality is certainly not sufficient. For instance, it was mentioned that traffic information provided by various media (radio, TV) is repeatedly not updated with sufficient frequency because the channel from source to driver is too long. The collection of the information is often subjective while different sources may provide conflicting information. These issues make the information in general not very reliable and accurate. Therefore, there would be scope for user-oriented and improved information systems.

Relative role of transport telematics in the range of traffic management tools

Transport telematics falls in a range of various measures to an improved road traffic management. Different measures are complementary to each other. Information provision to drivers in various forms will probably play a strategic leading role, since it has relatively few drawbacks.

Although it was expected that debiting systems have a relatively high potential to smooth traffic flows in relation to the potential for the provision of traffic information, this measure has a weak public support. Especially in the case of advanced automatic debiting systems which trace car trips, privacy aspects are a sensitive matter. The lack of public support is believed to be a serious bottle-neck to the implementation of debiting systems.

Respondents were not optimistic about the impacts of priority measures. This measure was regarded as a sub-optimal measure in the sense that it restricts a given corridor from full utilization of its total traffic flow capacity by the exclusive use by only a small number of vehicles. Another drawback may be the technical and operational problems (an example was given of a priority lane experiment on the A1 motorway in the Netherlands in 1994).

The disadvantages inherent to debiting systems and priority measures leave more scope for information tools as an alternative measure to meet traffic problems.

Optimal types of information and ways to present information

It was felt that there is only scope for driver information systems with simple interfaces. For instance, motorway traffic information by videotext, with a relatively complex interface, will never be consulted by a significant number of people.

From the different types of information which could be provided, most benefits are expected from the provision of improved dynamic information on congestion and obstructions on motorways. Efforts have been started also to develop an information system that is able to predict travel times on motorway segments. It is believed that information on expected travel times (for individual routes) is especially useful for user groups as business travellers.

A good option would be to let the driver select his own route choice; this might be as effective as providing complete route recommendations.

Basic requirements

Three basic requirements were mentioned by the respondents which are seen as absolute conditions for a successful implementation of transport telematics technology on motorways. These are first, the development of an open or flexible system architecture according to agreed standards, second, fully reliable information, and third, a full acceptance by road users.

While the first issue relates to the hardware of the technology, the latter two relate to the user side. It is recognized that user acceptance is of critical importance for the further development of transport telematics technologies. Almost all respondents mentioned the issue that when people are once mislead by false information, all belief in such a system will immediately decline.

Drawbacks
The use of transport telematics technologies also presents possible (negative) side-effects. The main side-effects are said to be the following:

1. Drivers may be distracted from the driving task by digesting too much information at one time or by a visual interface. These problems mainly concern in-car systems, and therefore there is more confidence in systems with speech interfaces. The size of text displays of variable message signs is also critical. Therefore, in most cases a limited number of characters in these messages is used.
2. In certain cases congestion may be not reduced but transferred to other parts of the motorway network (like 'squeezing a sausage').
3. Another problem is that an undesirable use might take place of secondary roads. However, opinions on the likeliness of this danger appear to vary. On the one hand one should recognize that this is an implicit impact of providing information on road obstructions. On the other hand there was not much fear that this would happen to a large extent. There is a large difference between different sites regarding this use of secondary roads. When the complete surrounding motorway network around a city is heavily congested, alternative routes have to be sought through the city itself. However, in cases where sufficient motorway capacity is available (e.g. when an urban ringroad is only partially congested, leaving the alternative to access or pass the city from the other side), this problem is smaller, especially when information systems aim to use this excess capacity (see our case study in Chapter 5).

In this respect, the importance of implementing ramp metering was emphasized, since this measure may reduce abuse of long motorway ramps and parallel roads.

Relationship between road managers and transport telematics industry
The supply of road management technology is said to be technology-led; it is believed that there are no technological barriers even in the short term. This supply market has a 'wild' and uncoordinated character, and is oriented on stand-alone standard products.
A primary task of the road manager is the implementation of an integrated road management system consisting of various well-interconnected telematics applications. Therefore, the importance of the development of integrated applications is stressed as going beyond standard products. Given also the high quality requirements of the technology, only a few companies are able to satisfy these requirements.
At present in many cases functional requirements are defined by the (national) road managers, and appropriate hardware is sought for on the market in the form of developed products by the industry. Examples are multi-modal travel information and

RDS-TMC. In other cases companies themselves are taking initiatives, offering solutions to traffic problems by complete hardware packages suited for particular sites.

Bottlenecks to the implementation process

Respondents were asked which were the most serious bottlenecks to a rapid future implementation of telematics technologies, out of the range of technical problems, operational problems, low cost/benefit ratios, and problems in the field of policy and finance.

There was a uniform consensus that technological and operational problems would play no significant role in the future. This consensus was largely based on the good experience with transport telematics applications which are already operational. This opinion was also shared by those involved in the operational side. A clear example was given of the use of new transmission cable materials which have enlarged the capacity of digital communications significantly and which is a requirement for many telematics applications.

A strong feature of transport telematics is its relatively high cost/benefit ratio compared to that of road capacity enlargements. It is believed that the cost/benefit ratios of telematics applications may be socially acceptable. A concrete example was given of experiments with variable advisory speed displays. The reduction of social costs by occurring accidents on the respective corridors seemed to be larger than the costs of these systems. Safety is thus an aspect where clearly benefits can outweigh the costs of the telematics technology. More uncertain are the cost/benefit ratios of road-side systems redirecting drivers. Benefits are here subject to the attitude of the road user; costs of the equipment however are relatively low compared with the potential combined effect on all road users.

The available budgets of road managers for investments in transport telematics have rapidly grown since 1990, which is in most cases mainly a shift in resources from building new roads. At least in the country of the Netherlands, there are no severe budget restrictions foreseen in the near future regarding the development and implementation of inter-urban road-based telematics systems. It may be concluded that technology costs are not seen as a possible bottle-neck for at least a broader implementation of public driver information systems.

An important role is being played by traffic policies of cities, which have an influence on the implementation of transport telematics on motorways in urban areas, thus those sites of the motorway network where the potential of transport telematics is at largest. Especially in Europe, many large municipalities have a transport policy aimed at reducing car use in favour of public transport use.

Although the argument plays that solely the availability of better travel information would not directly lead to more car use and would have no negative impact on the modal split, telematics technologies in general have the image of being car user-friendly, which clearly conflicts with local policy goals. These conflicting goals between local and national traffic management have led in the past to putting efforts in the development of driver information systems with explicit guidance to park and ride locations, which should also stimulate the use of public transport by telematics.

There are expectations that in the near future the barrier of resistance of large cities to a wide-scale implementation of transport telematics can largely be overcome. One may argue that given the fact that economic development makes mobility, including car mobility, a necessity, and either roads have to be used more efficiently or new road capacity has to be created, telematics would be a preferable means to increase capacity. Furthermore, local political resistance is always temporary and dependent of city councils. Transport telematics is seen as a new phenomenon, and the public acceptance process has only recently started. The developments of various telematics tools are market determined processes and given these (autonomous) processes, the best option would be to harmonize their development as much as possible in line with the objectives of national road management as well as those of cities. In short, it is believed that local policy can only slow down the implementation process but that it will not be an remaining bottleneck.

Parallel existence of various systems

The super-implementation of new systems on other (existing) systems is an inevitable aspect in dynamic road transport management. A technical potential danger could exist in conflicting information provided by the various systems. The size of this danger may however not be large when one single monitoring system is used to collect floating car data and other kinds of information. The use of this one system could potentially ensure that various applications use the same source data (given that there will be no significant organizational problems), by which the chance of conflicting information outputs is reduced. However, problems might still occur when the various systems have different processing times, but it is expected that these differences will be reduced in the near future.

The consistency in traffic information provided to the end-user is seen as a primary task of the public authorities. Although technical improvements will facilitate the achievement of this, much attention is still needed. It is one of the main reasons for the involvement of road managers in the development of in-car information systems, products that are typically for the private consumer market.

Differences in the potential of telematics between different regions

An important issue is the difference in the potential of transport telematics between regions with different characteristics of road networks. It was generally assumed that this potential is higher for densely crowded areas, with high road densities and average to high levels of traffic congestion (see also our case study in Chapter 4). It is evident that in these dense regions (which cover large parts of the Netherlands and Germany), there would be less scope for additional road capacity extensions than in other regions due to a lack of space. This characteristic would strengthen the relative need for transport telematics technologies. A clear example may be given of the situation in Hong Kong. In this region the development of new transport systems and telematics was accelerated by the high population concentration.

Some arguments may however also be placed for the reverse of this assumption. First, in some regions due to the highly crowded situations, the capacities of the complete motorway networks are used to their critical limit. Transport telematics would then in many cases transfer congestion problems instead of solving them. The second argument was that in such urban regions the motorway networks fulfil a dominant role for regional traffic beside international traffic, due to the high density of the motorway network. Secondary roads then usually offer no good alternative, and this contrasts with regions with low motorway densities (e.g. in a country like France, the Routes Nationales offer more scope for rerouting motorway users). A third reason that can be mentioned here is that the relatively short average trip distances of regional motorway users make drivers sensitive to short detour distances, and this means that they would be more reluctant to make route changes.

Implementation prospects

In the past five years the development of transport telematics systems has shown a rapid acceleration. The driving factor behind this has been the growing public census that unlimited expansion of road infrastructure will put serious damage to the environment and will not be sufficient to meet all kinds of traffic problems. However, until recently the emphasis has mainly been on the development of the technology itself and pilot tests. It is agreed that now is the time for broader implementation of those systems that are fully standardized. It is expected that various different systems may exist parallel to each other, given the fact that the end-user market will be segmented.

It is believed that the actual rate of development of transport telematics technologies will continue for at least fifteen years. Although it may show development waves dependent on the level of public interest, in the longer run it is expected that no limits will exist regarding the development of new generations of transport telematics systems.

9.5 Conclusions

Transport telematics are said to be able to significantly contribute to the achievement of two of the three main policy objectives of transport telematics, namely traffic safety and traffic efficiency. As expressed in this chapter in an in-depth interview series, from the point of view of the public sector, in situ road managers, strong success factors are believed to be their relative low costs and the existence of only a few drawbacks in relation to other means to improve traffic performance. However, there are still some drawbacks in the case of disseminating driver information, like a possible distraction from the driver's task by the continuous availability of this information, and a potential (undesirable) use of secondary roads.

Major potential bottlenecks, which might delay the process of implementation of inter-urban telematics applications, will probably not lie in the fields of the technological possibilities, the operational context or finance of the technologies. It is often believed that conflicts with local (car unfriendly) policies of large cities may cause some delays in this development, but this would not obstruct the implementation in the long term.

Based on the developments in the past five years, it is expected that the actual speed of transport telematics developments will continue at a similar pace, at least in the next 15 years. It is expected that in the longer run no limits will exist to the development of newer generations of transport telematics systems. This again stresses the need for pan-European standardization.

It is important to discuss with decision makers their views on the possibilities and potential for transport telematics in the short term and the key role that they should play. The market will react in the short term where commercial opportunities arise, particularly with respect to in-car systems and freight management systems (e.g. Electronic Data Interchange). Here there are clear commercial incentives to be at the leading edge of transport technology, and the costs can either be passed on to the final consumer or be seen as a saving in the operating costs of a firm or employee. However, it is in the public sector where most action needs to be taken. It is here that most of the research and development will take place. It is here that the necessary legislative charges will have to be made. It is here that most of the finance will be made available. It is here that most experimental sites will be set up to test the various technologies. And it is here that standards will eventually be set. The public sector has the crucial role in determining the scale, nature and range of transport telematics implementation.

In addition to these multifarious roles for the public sector, there is a further responsibility that they have. This is the communication of the technological opportunities to the public. Underlying much of the empirical evidence cited here from the various case studies and the in depth discussions with decision makers is a realization of the potential for transport telematics. However, that opportunity has not yet been translated into action. There is widespread interest from industry and the general public, but there is also a healthy scepticism about the actual usefulness of the technology. Many users are prepared to try the technology and to give a view on it, but this often only relates to a limited, often personal experience. Much of the technology is changing the system at the margin. It is evolutionary, not revolutionary. Even where the systems are provided free at the point of use, opinions vary as to its usefulness. When payment is suggested, the reaction becomes neutral or negative. The greatest challenge to the public authorities may not be in overcoming the technical, financial and legal difficulties, but in convincing industry and the general public of the real benefits of transport telematics.

10 Conclusions and Policy Recommendations

10.1 Synthesis and Purpose

Currently, there is a revolution taking place in the provision of transport services, and it affects the production and design of vehicles, road traffic operating systems, information systems, and the operation and management strategies of firms. In the longer term, there are visions of the intelligent vehicle and the automated highway. The basic premise underlying this telematics revolution is that through the use of technology, we can improve the efficiency, the environmental quality and the safety of our transport systems, and reduce the levels of congestion, without actually increasing the amount of physical road space available. It seems that this is a no loss situation as society, firms and individuals will all gain from the increased use of transport telematics.

This book has taken one part of this debate and presents an in-depth analysis of user response to these new telematics technologies. The basic argument promoted here is that it is only through an understanding of the behavioural responses to innovation and technology that its impact can actually be measured. Furthermore, this understanding sets the conditions for optimal implementation of the range of systems available, so that acceptance is maximized and the market potential is fully realized.

The approach taken has been to identify the main actors and to match these with the applications of telematics where user response can be expected. We then systematically presented data for each of these applications from a range of European case study locations. The empirical analysis, together with the descriptive statistics, was complemented by a more technical multivariate analysis. The novelty of the book is in bringing together of the case study material and the presentation of the data within the actor-based framework. This covers the supply of telematics, the demand for telematics, intermediate actors and the external actors in the process (Chapter 3).

In addition, this book redresses the balance. Too often there are optimistic claims made about the way in which innovation and new ideas will change lifestyles. For example, there was a clear argument that telecommunications systems would replace the need to travel, and that many people could now be working from home. Reality has proved more complex and it is now suggested that telecommunications systems act as a complement to travel decisions and may even increase the demand for travel (Salomon

1990). Change is much more subtle, and the Nested Approach together with the Cascade System present the range of different responses and the processes through which change takes place. Again, the case study material amply illustrates the range of responses from actual applications. The underlying message is that we must understand, measure and evaluate behaviour before we can make optimistic claims about the impact of the technology.

Nevertheless, transport telematics, as an innovative technology, must have a strong beneficial impact on society. Basically, this covers three kinds of benefits: collective (societal) benefits, commercial benefits and individual (private) benefits. From the policy perspective of directing new technologies, transport telematics is a complex issue. It involves several distinct application areas which each have different actors, and they have different impacts on society. The ultimate goal of policy makers is to reach the maximum welfare level which is achievable given the possibilities of a new society-wide technology. Such a maximum welfare levels refers to a situation in which the highest possible benefits are achieved for the whole society in terms of road safety, traffic efficiency and a reduced pressure on the environment.

This raises the question as to what the first-best policy for the development of transport telematics would be. If the implementation is left to the market, the full potential may not be realized as actions would also be needed from the national and international agencies. For example, the setting of standards and necessary enabling legislation. If the implementation is left to the public authority, then again the full potential may not be realized, as innovation, marketing and promotion is more dynamic in the private sector. It is likely that the first-best strategy will depend on a combination of actions from both sectors with the exact mixture also being dependent upon the particular application under consideration. The role for research would be to establish the appropriate conditions and combination of private and public sector involvement.

This book has been selective in its focus on particular situations where user response can be expected. Even if first-best policies can be identified, that situation may change as technological innovation is continuous with improvements taking place all the time. Similarly, the scope is selective and not comprehensive. The enormous range of potential applications have been summarized in Chapter 2, and as noted above, we have been selective to three main areas of application in urban and inter-urban locations.

10.2 The Main Areas of Application - The Case Studies

10.2.1 Travel and Traffic Information Systems

Travel and traffic information systems comprise a wide range of information facilities for travellers. This book has gone into empirical depth concerning the behaviour of car drivers in response to the provision of (public) traffic information through variable message signs. Such systems are becoming standardized in the traffic environment in several countries. Their uniform character (they have similar displays and provide information about current traffic conditions, mainly about the occurrence of traffic jams) provides a sound basis for the measurement of drivers' responses to information, and to draw wider inferences from the individual pilot studies.

It has appeared, that for a given site, a significant share of passing vehicle drivers have positive attitudes towards dynamic traffic information, with a significant share of them actually deciding to reroute in particular situations. This finding confirms that from the individual road user's perspective, that there is indeed a need for such a kind of information, although needs may vary between different sorts of information, the different locations, the different situations and for different users.

A factor of importance for the development of traffic information systems is whether these systems will be public or private, since this affects the basic system functions and configurations. In both cases, they have to be integrated into the broader traffic management schemes, suggesting that the intervention of the public sector is essential. This will require a coordinated management in which public and private sectors must cooperate closely in the areas of traffic data collection and the dissemination of information. The complexity of management will be higher in the case of the individual systems. On the other hand, individual systems can be more effective, since they provide information which is specifically needed in individual situations. Therefore, an important policy choice will be whether emphasis should be given to the development of individual or collective systems. Apart from the management aspects, both kinds of systems also have different financing implications. Empirical evidence shows that the willingness to pay for having dynamic traffic information continuously available in the car is rather low, which suggests that the market for these systems could be lower than sometimes expected. In such a case, there would be an incentive for public authorities to supply public telematics services to achieve the desired traffic levels.

10.2.2 Public Transport Information Systems

Complementary to the information systems for drivers is a similar system public transport users. Detailed analysis has been carried out on systems in Southampton and Birmingham where real time information is presented to bus users about the next bus arrivals at the bus stops. The general expectation that real time information will generate additional trips (and revenue) has not been realized - the best estimate is a 5% increase in trips. However, there have been a large number of new bus users on the routes, and many bus users have a car available as a driver or a passenger. These findings suggest that the bus market is not as 'fixed' (or captive) as has been assumed in the past, but that it is much more 'fluid' in its nature.

The main impact of the bus information systems has been in the perceived quality of bus services, particularly in terms of its reliability, even though there are still concerns over the accuracy of the information provided. It is important to maintain this improvement in the quality of the bus service as it helps to retain the loyalty of existing and new users. When monitoring the changes in behaviour resulting from the introduction of bus information systems, frequent bus users and new users should be the main disaggregations used.

In terms of travel behaviour, the responses to bus information systems are diverse. The greater flexibility in travel patterns permitted by bus information systems has allowed users to switch to another route or destination, walk all the way or go to a shop and come back to the stop. High quality information gives the bus user greater control over their actions. The young and those in employment are most positive in their response to bus information systems, and they are the groups most likely to adapt their journey as a result of the real time information.

As with car-based information systems, the main actor is the public authority in conjunction with the bus operator. Successful implementation has taken place where public transport information systems have been combined with other means to promote public transport (e.g. priority at signalized junctions or exclusive rights of way for buses). Partnerships have been established between the private sector suppliers of the information displays, the automatic vehicle location devices, the roadside beacons and the software, and the local authority for the financial resources and opportunity to test the hardware and the software. The co-operation of the bus companies has also been assured. These three groups of actors have ensured a successful implementation.

The actual behavioural responses have been different to those expected in advance or even used to justify the introduction of the telematics technology in the first place. Further marketing and identification of new bus users (particularly the young and housewives, or those returning to work) is required so that their awareness of the system can be raised. The monitoring and evaluation of the public transport information system needs to take place over an extended period of time so that a clear understanding of the full range of behavioural responses can be made.

10.2.3 Freight and Fleet Management

The telematics applications in the field of freight and fleet management are in general different to other application fields, as there is a strong economic incentive for these systems. The increasing complexity of physical logistics in road freight transport requires a parallel circulation of information by advanced communications systems between the various logistic agents involved (co-operative communication), and between the logistic operator and trucks on the road (internal communication). Telematics applications that are being developed in the field of logistic freight management concern applications like electronic data interchange between transport agents, as well as automatic vehicle location systems and mobile communications systems for which there are various technological options. It is clear that there is a huge industrial potential for these technologies.

The take-up of these technologies seems to have been slowed down for a number of reasons which may be regarded as bottlenecks to a wide-scale implementation. These bottlenecks are found in the software of some applications as well as in the financing and organizational aspects of the applications. First, the applications may not correspond with the professionals' expectations. Large operators possess skills which allow them to dominate these applications, notably for the installation of very efficient private telecommunications systems. In the case of some technologies which may assist truck drivers in their navigation task, there is a general resistance based on the fact that truck drivers are often familiar with the routes they drive. This kind of resistance is a direct result of the limitations of the software being used.

Secondly, there is the issue of finance. Only a few firms, probably the large and international firms, show real interest in the use of advanced (global) information systems. In the freight sector, which is largely made up of small and medium sized companies, the scale of such investments necessary for access to this information means that only the larger companies can compete. This has implications for the smaller companies as their competitive position may be weakened. Regulating policies may be necessary to direct the implementation process of road freight telematics in a more stable direction so that there is no unfair competition between the different types of operators.

Thirdly, there is the organizational aspect, which is extremely important for the implementation of electronic data interchanges systems. Although standards are well developed for these systems, the use and further take-up of such systems are mutually dependent on the number of subscribers. A policy aimed at promoting the use of electronic data facilities could be very effective through the 'snow-ball' effect initiated, which will result in a faster and more widespread adoption of the technology, with prices being kept at a lower level.

For the road freight market, the transport telematics and the mobile telematics systems are not yet a common feature. A more advanced logistic management system involving the use of transport telematics would favour the larger companies. It is clear that although the developments in the freight sector can be promoted primarily by the private sector, there is a strong case for complementary actions by the public sector to standardize systems, promote the potential of telematics (particularly among small firms), and to facilitate the change to new more efficient operating systems, through training, information and possible tax benefits.

10.2.4 Conclusions on the Case Studies

From the case study material presented in Chapters 5, 6, 7 and 8, certain interesting conclusions can be drawn:

- Behaviour depends on a wide range of variables, including strategic issues such as the expected network conditions, tactical adjustments, relating to the actual conditions met, and temporal disequilibrium or a general inertia towards change. It is difficult to predict the conditions under which user response might change.

- Travel decisions are not only based on time minimization, but the desire to encounter minimum levels of congestion and to arrive at ones destination on time. Familiarity with and knowledge of the road system or the public transport network are key determinants of the willingness or the ability to change behaviour. Many users of the road and public transport systems do not have a sufficient level of knowledge to change their behaviour and they are cautious in their acceptance of information given to them.

- There is a complementarity between information systems and behavioural responses. Pre-trip information allows more options for change including departure time changes, mode switching, or suppressing the trip altogether. In-trip information restricts changes to route, destination and possibly mode.

- It seems that certain thresholds need to be exceeded before behavioural change takes place, and this varies by situation and respondent. For example, a bus user might only switch mode or route if the next appropriate bus is not due to arrive for 10 minutes, or a car driver might switch routes if the queue length on the current route exceeds 5 kilometres. Further research is required to establish the scale and nature of these thresholds.

- There is a strong resentment against having to pay for information. Those involved in business travel have a higher willingness to pay, but this may only be a transfer payment from the employer to the charging authority. This finding reflects similar conclusions carried out on user response to road pricing systems.

- Telematics has aroused a generally positive response, particularly from the young and those in employment, mainly because their overall travel time and travel time uncertainty can be reduced. This positive attitude needs to be built upon and strengthened through promotion and marketing, from the local authorities and private sector operators and companies.

- Natural markets do not exist and there is a substantial marketing role to be filled. Many applications depend on local or site specific conditions which may be decisive for the market potential of telematics systems. The benefits from real time information needs to be clearly marketed to individuals and firms in a personalized way - it needs to be related to their particular circumstances and location.

- For many freight companies, the initial potential seems greatest in electronic data interchange systems, not in more sophisticated telematics applications. They need to be convinced of the competitive advantage of having (or the competitive disadvantage of not having) the technology, and the impact that it will have on their operating costs. Only then will they invest in the skills, the hardware and the software.

As with all innovations, there is no certainty that any particular application will be successful, yet the potential for transport telematics has not yet been fully exploited. Delays in the understanding of behavioural response, and the appropriate ways to identify market opportunities, may result in the potential of many innovations not being fully realized.

10.3 Policy Perspectives

Transport is now entering a new era. Concerns have switched from large scale investment in new infrastructure, at least at the city level, to limiting the negative effects of transport - the externalities. Transport causes congestion, pollution and accidents, and technological options are now seen as a means to reduce these external costs. Informatics and telecommunications have caused a revolution in data handling, processing and transmission, and this in turn has impacted on all households and firms.

The new transport telematics, and the information and communications technology more generally, have already had a profound impact on the mobility of people and the movement of goods, and their impact on the relative competitive advantages of particular regions is likely to further increase their importance (Capello 1994). This technology may now have an important role to play in reducing levels of congestion and improving the environment through better management of the road system and better information to the user. The technological revolution is the third great change in modern society, following on from the agricultural revolution in the 18th century and the industrial revolution of the 19th century.

Transport telematics is directed at improving the performance of the transport sector by a variety of technological innovations focusing on behavioural changes of travellers as a result of a better provision of information. Telematics has become a widely researched topic, in Europe (EU DRIVE and PROMETHEUS programmes), Japan (the VICS programme) and the USA (the IVHS programme). Technology is regarded as one means to alleviate the stress on a mobile society resulting from inefficient road use, to reduce the level of accidents and to limit the environmental damage caused by high pollution levels (Banister 1994a and b; Kawashima and Fujii 1992). Responses and research priorities in the different continents has varied tremendously as the congestion and environmental induced imperatives vary, and as there are fundamental cultural differences in the readiness to adopt technological solutions.

However, a significant improvement in traffic performance can only be reached if there is a substantial user response to new forms of information supply, user charging mechanisms and debiting systems for road users. Behavioural studies in the past show ambiguous results on this subject (e.g. EURONETT 1992; Mahmassani and Herman 1990; Mannering et al. 1994; and Ng et al. 1995). The need for a thorough assessment of the user side is therefore increasingly recognized in transport policy as behavioural responses depend heavily on the technology used (cf. Catling and Keller 1993; Klijnhout 1990; and Smith 1991). These have been summarized in Chapter 3 (Table 3.3).

More recently, the impacts of telematics, and information and communications technology more generally, have pervaded much of our everyday life, but the impacts have not been uniform, nor is there a strict cause and effect relationship. There is a tremendous potential for further developments in technology which can affect all aspects of peoples lifestyles, and the way in which business and industry is carried out. Yet there is also a reluctance among users to accept the new technology without question. It is here that the research reported in this book has a major importance. Unless the gap between the technologically soluble problems of transport telematics and the user acceptability can be bridged, then the potential opportunities and markets are greatly diminished.

Three main policy issues have emerged:

1. *Behavioural responses* are much more complex than originally thought. To some extent, the actor based approach outlined in Chapter 3 captured this complexity, but it seems that behaviour is neither utility-based nor satisficing-based. It is not clear whether time minimization or other factors, such as information reliability and familiarity with the road system or the public transport system, may be more

important. Both the interpretation of the advice given and the adherence to that advice means that the range of responses is great and the scale of change is often quite small. Of equal importance is the quality of the information and the way it is presented to the user. As with many innovations, there is a curiosity utility which encourages individuals to try out the system. The success or failure of a particular system is based on one or two good or bad experiences of it. Alternatively, it could be argued that it is more important to spend resources on the system itself so that it can be made more reliable. The irony here is that telematics may have a role to play that is inversely proportional to the reliability of the system for which the information is required. A completely reliable system does not require transport telematics or real time information.

A second major conclusion is that different people and firms respond in different ways to the same information, and that some are more amenable to trying out the new technology. There are important equity implications arising from the technology and this leads to greater benefits for the technologically literate and those willing to try out the technology. Safeguards must be introduced to ensure that there are no losers from telematics. All of these issues merit further investigation.

2. *Information* is a valuable commodity and questions have been raised in this book about the ownership, availability and access to that information. All of the telematics systems covered here are dependent on the accuracy and immediacy of the information provided. Undermining either of these characteristics results in a reduced value for the information. Much of the information is collected from the roadside by the public sector and is deemed to be in the public domain. However, there are costs associated with its collection, so commercial organizations might be expected to be charged for its use. If this is the case, then they also have to be given the opportunity to recoup their costs from the user. But willingness to pay is low. This dilemma has not yet been resolved in either the driver or the public transport information applications. It is only in the freight sector that a market for information has been established through data interchange and commercial computer networks. There must be economies of scale in information collection and distribution, so there may be a clear market opportunity for new information providers. But even here, safeguards will be needed.

3. *Public authorities* have the major role to play in the transport telematics sector. This instrumental role must operate at all levels of decision making, as the economic and social benefits of the technologies require a commonality of planning, regulation and harmonization for a successful implementation. In addition, there is a clear need for financial support from local authorities, and their expertise in promotion, publicity and awareness raising is also crucial to successful implementation.

National and local governments should therefore take the lead and ensure a minimum level of duplication in developments of in-vehicle technologies and infrastructure, while economies of scale will be achieved through the harmonization of functional requirements. In addition, there is a wide range of different technologies already in use and being developed in the specific field of operational road management. The implementation of such technologies can in principle be controlled more centrally than is in the case for the other telematics application areas. In the long run, the ultimate goal of the policy by public authorities will be an integrated traffic environment in which private, commercial and public telematics services will be available on a national and international scale.

A strong and critical role of the public sector lies in the software side of transport telematics applications. Public road managers traditionally have had the task to monitor traffic conditions and road infrastructure in order to determine longer-term developments in traffic flows. The collection of data feeds into most dynamic traffic information systems, public as well as private. However, some information technologies will be able to make use of their own data sources (e.g. through bi-directional communication between vehicles and operator centres). The means by which the public sector manages the source information and the possible contradictions between systems using different source data will be a principle factor determining the final organizational structure for the supply of telematics services.

The high expectations of the potential of transport telematics systems may overestimate the real impact, and it must be recognized that its direct contribution to the reduction in the social costs of transport may be modest. However, the dynamic generative capacity of transport telematics may nevertheless be substantial, especially when the technologies become standardized and widely applied. Only then will the true welfare and efficiency gains be realized.

Clearly, there is a huge potential for transport telematics systems, but it may not have the expected impacts. Rather than improve the efficiency and reduce the external costs of transport, it may actually increase levels of traffic congestion and trip lengths across the road network as a whole for the following reasons:

- There is a potential for overreaction to the information if too many individuals are aware of the predicted problem situation;

- There is the question of the optimum penetration levels, bearing in mind that there are many different types of transport telematics, but users will only select one. The systems do not directly compete with each other, but they are complementary;

- The nature of the message is also important as it is linked with adherence to advice. Minimal messages are ignored and a fuller explanation increases adherence to advice;

- User responses are likely to be complex and new methods may be needed to explore these changes. It will provide greater flexibility in trip making and link in closely with the user's own knowledge base;

- Pre-trip information may be more useful than in trip information as a greater range of choices can be identified. Once a trip is started, the range of alternatives is reduced to route, destination and possibly mode.

The performance of the transport system in Europe is not optimal, as its efficient operation is hampered by a wide range of external costs. The attraction of telematics is that reductions in the social costs of transport can take place without any increase in the prices charged. This option is acceptable to both politicians and the public. If it can be seen that through the use of transport telematics, reductions take place in the levels of congestion, pollution and accidents, then there is likely to be widespread support for the technology. Subsequently, it may be necessary to charge for the use of the technology, either directly to the users or indirectly through public agencies, but the acceptability of charging is likely to be much higher provided that tangible benefits can be observed. As with all new technology, the public need convincing that it actually works to their advantage. The demonstration effects of transport telematics needs to be widely publicized so that the most efficient use of the existing transport and telecommunications infrastructures can be achieved. Once this transition has been achieved, then the commercial potential of transport telematics can be fully realized.

The challenge facing decision makers in the 1990s is to convince people that the use of transport telematics and charging for real time information and the use of road space is of benefit to both the individual and society as a whole. The problems created by the advent of mass car ownership resulted from the clear benefits to the individual, but at a tremendous cost to society in terms of the externalities caused. The opportunity has now arisen to redress that balance, but the question still remains as to whether telematics will fulfil the expectations held of it and whether people are prepared to accept that challenge.

References

Al-Deek H, Martello M, May A and W Sanders (1989) Potential Benefits of In-vehicle Information Systems in a Real Life Freeway Corridor Under Recurring and Incident Induced Congestion. 1st VNIS Conference Toronto

Argyrakos G, Giaoutzi M and K Petrakis (1994) Travel Behaviour and the Diffusion of New Technologies. Proceedings 7th International Conference on Travel Behaviour. Valle Nevado, Santiago Chile 771-785

Association of Electronic Technology for Automobile Traffic and Driving (1992) Main Results of Research and Investigation on Super Smart Vehicle System. JSK International Forum. Document no.3. Tokyo

Associazione Italiana Società Concessionarie Autostrade e Trafori (AISCAT) (1992) Informazioni. no 3-4/92. Rome

Banister D, Nijkamp P, Camara P and G Pepping (1994) The Analysis of User Response to Advanced Transport Telematics - Measurement, Methodological and Conceptual Issues. BATT internal working paper. DRIVE II Programme. EC DG XIII Brussels

Banister D, Capello R and P Nijkamp (Eds) (1995) European Transport Network Policies. Belhaven London

Banister D (1994a) Road Transport Informatics and Road Transport Developments. In: Nijkamp P (Ed) New Borders and Old Barriers in Spatial Development. Avebury Aldershot UK 229-245

Banister D (1994b) Equity and Public Acceptability Issues. In: Internalising the Social Costs of Transport. OECD Paris 113-152

Banister D (1995) Telematics and Road Transport Built Environment. vol 21 no.4, (forthcoming)

Batten D, Carti J and R Thord (Eds) (1994) Networks in Action. Springer-Verlag Berlin

BATT (1992) Identification of Requirements. Deliverable 2. DRIVE II Programme. EC DG XIII Brussels

BATT (1994) Assessment of Behavioural Impact at the Inter-urban Level. Deliverable 13. DRIVE II Programme. EC DG XIII Brussels

BATT (1995a) Results of the Southampton After Survey and Comparison with the Before Surveys. Deliverable 11A. DRIVE II Programme. EC DG XIII Brussels

BATT (1995b) The Impact of Providing Real Time Passenger Information at Bus Stops: Results of the Birmingham After Survey and Comparison with the Before Survey. Deliverable 11B. DRIVE II Programme. EC DG XIII Brussels

Ben Akiva M, Bergman M, Daly A and R Ramaswamy (1984) Modelling Inter Urban Route Choice Behaviour. In: Volumuller J and R Hammerslag (Eds) Proceedings 9th International Symposium on Transportation and Traffic Theory. VNU Science Press Utrecht The Netherlands

Berkum E van and P van der Mede (1993) The Impact of Traffic Information. PhD Thesis. Universiteitsdrukkerij Delft The Netherlands

Bertuglia C, Fischer M and G Pieto (Eds) (1995) Technological Change, Economic Development and Space Springer-Verlag Berlin

Blaas E, Vleugel J, Louw E and T Rooijers (1992) Autobezit, Autogebruik en Rijgedrag. In Series: Infrastructuur, Transport en Logistiek. Delft University Press Delft, The Netherlands

Bollo D (1992) Development of New Information Systems for Freight Transport Operations in Europe. IATSS Research. vol 16, no 2 57-69

Bonsall P and A May (1986) Route Choice in Congested Urban Networks. In: Research for Tomorrows' Transport Requirements. Proceedings WCTR Vancouver. UBC Vancouver BC Canada 1407-1425

Bonsall P (1992) The Influence of Route Guidance Advice on Route Choice in Urban Networks. Transportation vol 19, no 1 1-23

Bovy P, Orfeuil J and I Salomon (1993) A Billion Trips a Day: Mobility Patterns of the Europeans. In: Nijkamp P (Ed) Europe on the Move. Avebury Aldershot UK 67-93

Bruinsma F, Pepping G and P Rietveld (1995) Infrastructure and Urban Development; the Case of the Amsterdam Orbital Motorway. In: Batten D and C Karlston (Eds) Infrastructure and Spatial Development. Springer-Verlag Berlin (forthcoming)

Buijn H and W Schouten (1994) Route Choice Information Amsterdam. Proceedings 27th ISATA Conference. Aachen 143-150

Button K, Nijkamp P and H Priemus (Eds) (1996) European Transport Networks: Concepts, Analysis, Policies. Edward Elgar London (forthcoming)

Capello R and A Gillespie (1993) Transport, Communications and Spatial Organisation: Conceptual Framework and Future Trends. In: Nijkamp P (Ed) Europe on the Move, Avebury Aldershot UK 43-66

Capello R (1994) Spatial Economic Analysis of Telecommunications Network Externalities. Avebury Aldershot UK

Caplice C and H Mahmassani (1992) Aspects of Commuting Behavior: Preferred Arrival Time, Use of Information and Switching Propensity. Transportation Research. vol 26A no 5 409-418

CARGOES (1990a) Analysis of Stability Properties of Route Guidance Strategies. Deliverable 14. DRIVE II Programme. EC DG XIII Brussels

CARGOES (1990b) Drivers' Requirements for Route Guidance Advice. Deliverable 18. DRIVE II Programme. EC DG XIII Brussels

CARGOES (1990c) Drivers' Reaction to Route Guidance Advice. Deliverable 20. DRIVE II Programme. EC DG XIII Brussels

Catling I and H Keller (1993) The LLAMD Euro-project: Expected Impacts of Dynamic Route Guidance Systems in London, Amsterdam and Munich. In: Advanced Transport Telematics. Proceedings of the Technical Days. vol II EC Brussels 70-74

Celant A, Dematteis G, Fubini A and G Scaramellini (1992) Caratteri Generali e Dinamica Recente del Fenomeno Urbano in Italia. In: Dematteis G (Ed) Il Fenomeno Urbano in Italia: Interpretazioni, Prospettive, Politiche, Franco Angeli Milan

Christaller W (1993) Die Zentralen Orte in Süddeutschland. Fischer Jena

Cooper J (1992) Road Freight Transport and the Single European Market. IATSS Research. vol 16 no 1 53-59

CORD (1993) Recommended Definitions of ATT Subfunctions, Functions and Areas. Deliverable D003. DRIVE II Programme. EC DG XIII Brussels

Dematteis, G (1985) Controurbanizzazione e Strutture Urbane Reticolari. In: Bianco G and I Magnani (Eds) Sviluppo Multiregionale: Teorie, Metodi Problemi. Franco Angeli Milan

Dudek C, Stockton W and D Hatcher (1982) Real-time Freeway-to-freeway Diversion: the San Antonio Experience. In: Freeway Operations, Railroad-highway Grade Crossings, and Evaluating Highway Improvements. Transportation Research Record no 841 Transportation Research Board Washington DC 1982 1-14

EC Transport Telematics 1993 Annual Technical Report. STIG Programme. EC DG XIII Brussels

Emmerink R, Nijkamp P and P Rietveld (1992) The Role of Information in the Performance of Transport Networks. MARTA internal working paper 6. DRIVE II Programme. EC DG XIII Brussels

Emmerink R, Axhausen K, Nijkamp P and P Rietveld (1994a) The Economics of Motorist Information Systems Revisited. Transport Reviews vol 14 no 4 363-388

Emmerink R, Nijkamp P, Rietveld P and J van Ommeren (1994b) Variable Message Signs and Radio Traffic Information: An Integrated Empirical Analysis of Drivers' Route Choice Behaviour. TRACE discussion paper no 94-127 Tinbergen Institute Amsterdam

EURONETT (1992) Final Report. DRIVE I Programme. EC DG XIII Brussels

EUROSTAT (1991) Regional Data Base: Regio, Description of Contents. Office for Official Publications of the EC. EC Luxembourg

Filippi F, Della Site P and L Pickup (1994) Assessing Users' Attitudes and Potential Market for ATT. OECD TT3 Omiya Seminar on Advanced Road Transport Technologies. Omiya Japan

Fischer M (1993) Travel demand. In: Polak J and A Heertje (Eds) European Transport Economics. Blackwell Publishers Oxford 6-32

Hamer R, Kroes E and H van Ooststroom (1991) Teleworking in the Netherlands: An Evaluation of Changes in Travel Behaviour. Transportation. vol 18 no 4 1991 365-382

Handy S and P Mokhtarian (1995) Planning for Telecommuting: Measurement and Policy Issues. Journal of the American Planning Association. vol 61 no 1 99-111.

Henderson D, Koenig B and P Mokhtarian (1995) Using Travel Diary Data to Estimate the Emissions Impacts of Transportation Strategies: The Puget Sound Telecommuting Demonstration Project. Journal of the Air and Waste Management Association. June

Hilferink M, Nijkamp P, Pepping G and D Fabbri (1994) Assessing the Telematics Potential on Network Segments in Italy. Proceedings 27th ISATA Conference. Aachen

Hilferink M and R Rinzema (1995) TEGIS: Telematica Evaluatie Geografisch Informatiesysteem. Economisch Bulletin. vol 26 no 6 24-28

Himanen V and P Nijkamp (1992) Environmental Quality and Transport Policy in Europe. Transportation Research. vol 26A no 2 147-157

Hinloopen E, Nijkamp P and P Rietveld (1983) Qualitative Discrete Multiple Criteria Choice Models in Regional Planning. Regional Science and Urban Economics. vol 13 no 1 77-102

Hinloopen E and P Nijkamp (1990) Qualitative Multiple Criteria Choice Analysis: The Dominant Regime Method Quality and Quantity. vol 24 no 1 37-56

IVHS America Strategic Plan for Intelligent Vehicle-highway Systems in the United States. IVHS America report. Washington DC May

Jeffery D Driver (1990) Route Guidance Systems: State-of-the-art. In: Soekkha H, Bovy P, Drewe P and Jansen G (Eds) Telematics - Transportation and Spatial Development. VSP Utrecht The Netherlands 151-172

Johansson B, Karlsson C and L Westin (Eds) (1994) Patterns of a Network Economy. Springer-Verlag Berlin

Kawashima H (1990) Japanese Perspective of Driver Information Systems. Transportation vol 17 no.3 263-284

Kawashima H and H Fujii (1992) A Perspective of IVHS Related Activities in Japan (second version) Association of Electronic Technology for Automobile Traffic and Driving. Tokyo

Keeble D, Owens P and C Thompson (1982) Regional Accessibility and Economic Potential in the European Community. Regional Studies. vol 16 no.6 419-432

Khattak A, Schofer J and F Koppelman (1991) Influence of Various Factors on Commuters' Enroute Diversions in Response to Delay. TRB Conference. Paper 910336. Washington DC

Klijnhout J (1990) Telematics and Traffic Management. In: Soekkha H, Bovy P, Drewe P and G Jansen (Eds) (1990) Telematics - Transportation and Spatial Development. VSP Utrecht The Netherlands 135-150

Kokkota S (1993) Road Transport Informatics (RTI) and Drivers' Behaviour. In: Nijkamp P (Ed) Europe on the Move. Avebury Aldershot UK 157-171

Kurami K, Hattori A, Hosaka A and M Taniguchi (1991) Development of the Autonomous Personal Vehicle System. In: Denshi Tokyo no.30

LISB Schlussbericht. DRIVE I Programme. EC DG XIII Brussels 1991

Lotan T and H Koutsopoulos (1993) Models for Route Choice Behaviour in the Presence of Information Using Concepts from Fuzzy Set Theory and Approximate Reasoning. Transportation. vol 20 129-155

Mahmassani H and G Chang (1987) On Boundedly Rational User Equilibrium in Transportation Systems. Transportation Science. vol 21 no.2 89-99

Mahmassani H and R Herman (1990) Interactive Experiments for the Study of Tripmaker Behaviour Dynamics in Congested Commuting Systems. In: Jones P (Ed) Developments in Dynamic and Activity Based Approaches to Travel Analysis. Avebury Aldershot UK 272-298

Mannering F, Kim S, Barfield W and L Ng (1994) Statistical Analysis of Commuters' Route, Mode and Departure Time Flexibility. Transportation Research. vol 2C no.1 35-47

MARTA (1992) State of the Art in Inter-urban Transport Telematics. Deliverable 2. DRIVE II Programme. EC DG XIII Brussels

MARTA (1993a) The Cascade System: A Comprehensive Analytical Framework for Assessing the Market Potential of ATT. Deliverable 3A. DRIVE II Programme. EC DG XIII Brussels

MARTA (1993b) ATT Market Assessment Report 1. Deliverable 4. DRIVE II Programme. EC DG XIII Brussels

MARTA (1993c) Preliminary Typology of European Inter-urban Corridors. Deliverable 5. DRIVE II Programme. EC DG XIII Brussels

MARTA (1994a) ATT Requirements and Responses of Inter-urban Fleet Operators (Netherlands). Deliverable 11A. DRIVE II Programme. EC DG XIII Brussels

MARTA (1994b) ATT Requirements and Responses of Infrastructure Owners (Netherlands). Deliverable 12A. DRIVE II Programme. EC DG XIII Brussels

MARTA (1994c) Regional and Corridor-specific Differences in the ATT Potential: A Case Study of Italy. Deliverable 13. DRIVE II Programme. EC DG XIII Brussels

MARTA (1995) Analysis of Company Intelligence on ATT Inter-urban Investments in Europe. Deliverable 9. DRIVE II Programme. EC DG XIII Brussels

Martellato D, Nijkamp P and A Reggiani (1996) Measures and Measurement of Network Accessibility. In: Button K, Nijkamp P and H Priemus (Eds) European Transport Networks: Concepts, Analysis, Policies. Edward Elgar London (forthcoming)

Masser I, Wegener M and O Sviden (1992) The Geography of Europe's Future. Belhaven London

Mokhtarian P (1990) A Typology of Relationships Between Telecommunications and Transportation. Transportation Research. vol 24A no 3 231-242

Mokhtarian P, Handy S and I Salomon (1995) Methodological Issues in the Estimation of the Travel, Energy, and Air Quality Impacts of Telecommuting. Transportation Research. vol 29A no 4 283-302

Muscarà, C (Ed) (1978), Megalopoli Mediterranea. Franco Angeli Milan

Ng L, Barfield W and F Mannering (1995) A Survey Based Methodology to Determine Information Requirements for Advanced Traveller Information Systems. Transportation Research. vol 3C no 2 113-127

Nijkamp P (1979) Multidimensional Spatial Data and Decision Analysis. Wiley New York

Nijkamp P (Ed) (1994) New Borders and Old Barriers in Spatial Development. Avebury Aldershot UK

Nijkamp P, Pepping G and Reggiani R (1992) Some Examples of European Networks and Corridors. MARTA internal working paper 7. DRIVE II Programme. EC DG XIII Brussels

Nijkamp P, Reggiani A, Pepping G and G Lampugnani (1993) European Corridor Typology: Description of Methodology and Data Base. MARTA internal working paper 13. DRIVE II Programme EC DG XIII Brussels

Nijkamp P, Rietveld P and Voogd H (1990) Multiple Criteria Analysis in Physical Planning. Amsterdam

Nijkamp P and E Blaas (1994) Impact Assessment and Evaluation in Transport Planning. Kluwer Boston

Nilles J (1988) Traffic Reduction by Telecommuting: A Status Review and Selected Bibliography Transportation Research. vol 22A no 4 301-317

Nilles J (1991) Telecommuting and Urban Sprawl. Transportation. vol 18 no 4 411-432

O'Neill P (1990) The Role of Public Authorities. Transportation. vol 17 no 3 313-328

OECD (1988) Route Guidance and In-car Communication Systems. OECD Paris

OECD (1992) Advanced Logistics and Road Freight Transport. in Series: Road Transport Research. no 845769 OECD Paris

Outram V and E Thompson (1978) Drivers' Perceived Cost in Route Choice. Proceedings 6th PTRC Summer Annual Meeting. PTRC London 226-257

Ouwersloot H, Nijkamp P and G Pepping (1995a) Advanced Telematics for Travel Decisions - An Econometric Analysis of the Stopwatch Project in Southampton. TRACE discussion paper. no 95-180. Tinbergen Institute Amsterdam

Ouwersloot H, Pepping G and P Nijkamp 1995b) Telematics and Freight Transport. TRACE discussion paper. no 95-181. Tinbergen Institute Amsterdam

Owens, D (1980) Traffic Information Broadcasting: Driver Reaction to Two Kinds of Traffic Message, a Pilot Study. TRRL SR 603. TRRL. Crowthrone UK

Ran, B and D Boyce (1994) Dynamic Urban Transportation Network Models: Theory and Implications for Intelligent Vehicle-Highway Systems. Springer-Verlag Berlin

RECLUS (1989) Les Villes Européennes. Rapport pour la Datar Paris

Reggiani A, Lampugnani G, Nijkamp P and G Pepping (1993) European Corridor Characteristics: Some Typologies by Means of European Maps and Diagrams. MARTA internal working paper 12. DRIVE II Programme. EC DG XIII Brussels

Reggiani, A, G Lampugnani, P Nijkamp and G Pepping (1995) Towards a Typology of European Inter-urban Transport Corridors for Advanced Transport Telematics Applications, Journal of Transport Geography. vol 3, no 1, 53-67

Rijkswaterstaat (1992a) Dynamic Traffic Management in the Netherlands. Rijkswaterstaat. AVV Rotterdam

Rijkswaterstaat (1992b) Electronische Plaatsbepalingssystemen en EDI in het Goederenvervoer over de Weg. Ministry of Public Works and Water Management The Hague

Salomon I (1990) Telematics and Personal Travel Behaviour with Special Emphasis on Telecommuting and Teleshopping. In: Soekkha H, Bovy P, Drewe P and G Jansen (Eds) Telematics. Transportation and Spatial Development. VSP Utrecht The Netherlands, 67-89

Shefer D and P Rietveld (1994) Congestion and Safety on Highways; Towards an Analytical Model. TRACE discussion paper. no 94-92. Tinbergen Institute Amsterdam

Theil H (1967) Economics and Information Theory. North-Holland Amsterdam

Turner J, Dudek C and J Carvell (1978) Real time diversion of freeway traffic during maintenance operations (Abridgment). In: Road user information needs, pedestrian movement, and bicycle travel patterns. Transportation Research Record. no 683. Transportation Research Board. Washington DC 8-10

Ueberschaer M (1971) Choice of routes on urban networks for the journey to work, Highway Research Record. no 369. 228-238

Uetakaya K, Okamoto H and T Kawabata (1992) Advanced Traffic Control and Management System in Tokyo. Tokyo Metropolitan Police Department Tokyo

Verhoef E (1994) External Effects and Social Costs of Road Transport. Transportation Research. vol 28A no 4. 273-287

Verhoef E, Emmerink E, Nijkamp P and P Rietveld (1994) Information Provision; Flat and Fine Congestion Tolling and the Efficiency of Road Usage. TRACE discussion paper 94-157. Tinbergen Institute Amsterdam

Wootton J, Ness M and R Burton (1981) Improved direction signs and the benefits of road users. Traffic Engineering and Control. vol 5 no 22 264-268

Wright C (1976) Some characteristics of drivers' route choice in Westminster. Proceedings 4th PTRC Summer Annual Meeting. PTRC London

Yang H, Kitamura R, Jovanis P, Vaughn K and M Abdel-Aty (1993) Exploration of route choice behavior with advanced traveler information using neural network concepts. Transportation. vol 20 no 2. 199-223

Index

Associate Authors

Yorgos Argyrakos
Transport Environment Development Systems
Kondylaki Street 9
11141 Athens
Greece

David Banister
University College London
Bartlett School of Planning
22 Gordon Street
WC1H 0QB London
England

Paulo Camara
University College London
Bartlett School of Planning
22 Gordon Street
WC1H 0QB London
England

Richard Emmerink
Department of Economics
Free University
De Boelelaan 1105
1081 HV Amsterdam
The Netherlands

Francesco Filippi
University of Rome (La Sapienza)
Department of Hydraulics, Transport and Roads
Via Eudossiana 18
00184 Rome
Italy

Maria Giaoutzi
Department of Geography
National Technical University
Zographou Campus
Athens
Greece

Maarten Hilferink
Department of Economics
Free University
De Boelelaan 1105
1081 HV Amsterdam
The Netherlands

Francesca Kenny
Transport & Travel Research
16 Stanton Harcourt
OX8 1RJ Oxford
England

Massimo Marciani
T.M.T. Pragma
Via Salaria 290
00199 Rome
Italy

Peter Nijkamp
Department of Economics
Free University
De Boelelaan 1105
1081 HV Amsterdam
The Netherlands

Hans Ouwersloot
Department of Economics
Free University
De Boelelaan 1105
1081 HV Amsterdam
The Netherlands

Gerard Pepping
Department of Economics
Free University
De Boelelaan 1105
1081 HV Amsterdam
The Netherlands

Kostas Petrakis
Transport Environment Development Systems
Kondylaki Street 9
11141 Athens
Greece

Laurie Pickup
Transport & Travel Research
16 Stanton Harcourt
OX8 1RJ Oxford
England

Aura Reggiani
Department of Economics
University of Bologna
Piazza Scaravilla, 2
40126 Bologna
Italy

MIX
Papier aus verantwortungsvollen Quellen
Paper from responsible sources
FSC® C105338

If you have any concerns about our products,
you can contact us on
ProductSafety@springernature.com

In case Publisher is established outside the EU,
the EU authorized representative is:
Springer Nature Customer Service Center GmbH
Europaplatz 3, 69115 Heidelberg, Germany

Printed by Libri Plureos GmbH
in Hamburg, Germany